Urban Logistics in a Digital World

Marzena Kramarz • Katarzyna Dohn
Edyta Przybylska • Izabela Jonek-Kowalska

Urban Logistics in a Digital World

Smart Cities and Innovation

palgrave
macmillan

Marzena Kramarz
Silesian University of Technology
Gliwice, Poland

Katarzyna Dohn
Silesian University of Technology
Gliwice, Poland

Edyta Przybylska
Silesian University of Technology
Gliwice, Poland

Izabela Jonek-Kowalska
Silesian University of Technology
Gliwice, Poland

ISBN 978-3-031-12890-5   ISBN 978-3-031-12891-2   (eBook)
https://doi.org/10.1007/978-3-031-12891-2

This Palgrave Macmillan imprint is published by the registered company Springer Nature Switzerland AG. The registered company address is: Gewerbestrasse 11, 6330 Cham, Switzerland

Preface

Urban development is inextricably linked to the growing importance of logistical support. There is already a lot of research in this field, and model solutions are known for urban logistics in different phases of city development and dedicated to different types of cities (Taniguchi & Thompson, 2010). The changes in cities that allow them to be called smart do not necessarily generate more logistics needs, but certainly require the adaptation of logistics solutions to new challenges. The concept of smart cities is already well established in the literature of the last decade or so. It developed at the end of the twentieth century. Nowadays, the interest in its implementation is growing steadily, as evidenced by the increasing number of studies and publications in this area, as well as practical implementations of Smart City solutions. The first publications appeared in the SCOPUS database in 2010; however, a very large increase in publications combining logistics and Smart City issues can be observed in 2019 and 2020 (2015—29 publications and 2018—113 publications). The increase in research in this area indicates its importance and relevance. Discussions are undertaken on the mobility of residents, the application of innovative solutions in passenger and cargo transport, and the impact of transport on the quality of life of residents. The last year has seen an increase in the number of publications dealing with the issue of e-commerce in the context of the organisation of urban freight flows, as well as publications focusing on practical solutions, including blockchain technology. One of the most recent approaches to this issue, comprehensively presenting the mobility of people and cargo transport, is the business model proposed by Koutras et al. (2021). The authors focus their model on the problem of sustainable urban development. The approach proposed by the authors is in line with the research objectives of the proposed book. Equally relevant to this publication are the research results presented by Russo et al. (2022) and Petrungaro and Trecozzi (2020) on knowledge management at the local level of the urban transport system. The authors point to the strategic solutions included in the Regional Transport Plan.

Analysing the literature on the inclusion of logistics aspects in the Smart City concept, it can be noted that they most often refer to the problem of mobility of residents, including passenger transport, segregation, collection, and transport of waste, much less to the problem of cargo transport. The aim of the book is to comprehensively cover the issue of changes in logistics support of cities resulting from Smart City solutions. Taking into account the existing body of literature, in our research we have considered the movement of both people and cargo in the city.

The research process we carried out involved three main stages. In the first stage, we developed a theoretical construct of urban logistics maturity and proposed a way to assess Smart City solutions. Then we carried out a survey research on a sample of 280 Polish cities. The research focused on logistical aspects as well as on the application of smart solutions in cities. This part of the research made it possible not only to diagnose current logistics solutions in Polish cities, but also to assign them to different phases of logistic maturity of cities and to identify gaps in the logistics support provided so far.

The final stage was to analyse the relationship between the level of application of smart solutions in a city and the level of logistical maturity of the city. The methodology adopted in the book, which combines the level of city intelligence with the analysis of logistics solutions in cities, is a consequence of the research questions posed:

1. What factors build an urban logistics maturity?
2. How to assess the logistical maturity of a city?

3. How to study the level of intelligence of cities?
4. What are the relationships between a city's level of intelligence and its logistical maturity?
5. What impact does innovation have on an urban logistics maturity?

We tested the developed theoretical constructs as well as the adopted methodology on Polish cities. In doing so, we also asked ourselves specific research questions:

1. What levels of logistics maturity can be distinguished among Polish cities?
2. What level of advancement of smart solutions do Polish cities have?

The layout of the content presented in the publication is a consequence of the adopted research model. We started (Chap. 1) by introducing the concept of Smart City (SC). Smart City is a very complex issue, which is also indicated by the literature in this area. In a holistic approach, it analyses both IT issues (IT and ICT technologies) and economic, social, and environmental aspects. This is in line with the evolutionary development of the SC concept, which currently covers three generations of smart cities (1.0, 2.0, and 3.0) and the concept of successive helixes involving an increasingly broader group of stakeholders in the functioning of SC (city authorities + entrepreneurs + universities + local community + environmental organisations). In such a complex range of topics, logistics issues, including those related to cargo transport, are described relatively rarely, despite the fact that it is one of the key determinants of quality of life in smart cities. Mobility issues are separately assessed in all major international Smart City rankings (European Smart Cities Ranking; City in Motion Index; CITY Keys; Smart City PROFILES) alongside such major evaluation areas as economy, human capital, governance, and quality of life.

The aspect of positioning the city in terms of the advancement of smart solutions is crucial for the research adopted in the publication and for answering the research questions posed.

We have devoted the second chapter to urban logistics. In our view, logistics in cities cannot be treated selectively and refer only to the mobility of residents or goods flows. Other logistics tasks are also important, including waste issues and logistics hubs. The concept of logistical support for cities requires adaptation of the infrastructure and must be dealt with comprehensively in the long term. We, therefore, attribute great importance to including logistics aspects in city strategies. Logistics solutions in cities and a systematic approach to logistics aspects by city decision-makers are factors indicating an urban logistics maturity. The urban logistics maturity construct is discussed in Chap. 3 of the publication. In Chap. 4, we focus our attention on innovations in urban logistics and, taking into account the knowledge presented in Chaps. 1 and 2, analyse solutions in selected European cities. In Chap. 4, we discuss the methodology for investigating the level of logistical maturity of cities and the level of intelligence of cities. The results of surveys conducted on a sample of 280 cities in Poland are presented in Chap. 5. At that stage, the cities were classified in terms of both the level of application of intelligent solutions and the level of maturity. Chapter 6, concluding the work, contains the results of the analyses in terms of examining the relationship between the level of application of smart solutions in a city and the level of logistic maturity of a city.

Our publication is addressed to academics, city institutions, and city managers. By handing over our publication to you, we would like to express our hope that the subject matter and the way it has been presented will be of interest to you and will lead to discussion and new and interesting insights into smart cities and shaping the logistical maturity of cities.

Gliwice, Poland   Marzena Kramarz
Gliwice, Poland   Katarzyna Dohn
Gliwice, Poland   Edyta Przybylska
Gliwice, Poland   Izabela Jonek-Kowalska

Contents

List of Figures

List of Tables

1

Smart City: A Holistic Approach

Origin and Essence of the Smart City Concept

The Smart City (SC) concept is a relatively new issue considered on very interdisciplinary levels due to the systemic complexity of the entity, that is, the city. It emerged in the twentieth century and was a response to the growing role of cities in society and economy (Lom & Pribyl, 2020). The current intensive development of this concept is due to the following circumstances:

- increasing urbanisation, with 50% of the world's population now living in cities (Sulemana et al., 2019; Zhang et al., 2019),
- predictions that 70% of the world's population will live in cities by 2050,
- the need to improve and enhance the quality of life of the inhabitants of existing cities,
- creating new cities in which the Smart City concept and all technological, social, and the environmental solutions associated with it are taken into account already at the design stage.

© The Author(s), under exclusive license to Springer Nature Switzerland AG 2022
M. Kramarz et al., *Urban Logistics in a Digital World*,
https://doi.org/10.1007/978-3-031-12891-2_1

It follows from the above that the development of interest in the SC concept is the result of practical needs raised not only by SC developers and visionaries, but also by local communities living in or wishing to live in smart cities. It is also notable that national and regional authorities are now thinking about and implementing the SC concept as early as the urban design stage. An excellent example in this respect is China, which has planned to create as many as 154 cities built from scratch with SC-related infrastructural assumptions.

Defining SC is worth starting by looking at the city itself, without the adjective smart, which allows only some cities to stand out. The city itself is an organised social system (Manczak, 2014). The characteristics of this system are the population living in the city forming a local community, the territorial proximity of the inhabitants, the communication accessibility resulting from the concentration of housing and technical infrastructure, and the fact that the inhabitants of the city carry out industrial rather than agricultural activities, as is the case in the countryside (Malarski, 2000). Cities are therefore distinguished from the countryside primarily by the type of activity carried out by their inhabitants, as well as by the population density, which is much higher in urban areas. In the past, cities were also distinguished by the lifestyle of their inhabitants (Czornik, 2004). Currently, this feature is less important due to the progressive homogenisation of rural and urban lifestyles (Manczak, 2014). However, it should be added that, ultimately, whether a territorial unit is a city or a village in the modern world in most countries is determined by legal regulations on the basis of which the unit is granted the official status of a city (Brol et al., 1990). This process clearly takes into account the above-mentioned substantive distinctions identified both in the literature and in economic practice.

In defining and describing a Smart City, it is assumed that the necessary conditions for being a city as described above have been met, but thanks to some additional distinctions, the inhabitants of such cities can live better, and more comfortably, more fully than the inhabitants of a "normal" city. This is because at the core of the idea of SC lies the key assumption that SC is created and improved to continuously improve the life of its inhabitants. It is essential that this improvement should occur in all areas of urban existence, taking into account the dynamics of the human life cycle and preventing all forms of exclusion, from technological, through social to economic.

There are many different definitions of smart cities in the literature. Their overview is presented in Table 1.1.

The analysis of the definitions presented in Table 1.1 indicates several common characteristics of a Smart City. First of all, it is the use of modern technologies and innovative IT and ICT solutions that are key

Table 1.1 Overview of Smart City definitions

Authors	Definitions
Angelidou (2014)	"A conceptual urban development model based on the utilization of human, collective, and technological capital for the enhancement of development and prosperity in urban agglomerations"
Barrionuevo et al. (2012)	"Being a smart city means using all available technology and resources in an intelligent and coordinated manner to develop urban centres that are at once integrated, habitable, and sustainable"
Bowerman et al. (2000)	"A city that monitors and integrates the conditions of all of its critical infrastructure including roads, bridges, tunnels, rails, subways, airports, seaports, communications, water power, even major buildings, can better"
Chen (2010)	"Smart cities will take advantage of communications, and sensor capabilities are sewn into the cities' infrastructures to optimize electrical, transportation, and other logistical operations supporting daily life, thereby improving the quality of life for everyone"
Giffinger et al. (2007a, b)	"A city is well performed in a forward-looking way in economy, people, governance, mobility, environment, and living, built on the smart combination of endowments and activities of self-decisive, independent, and aware citizens"
Hall (2000)	"A city that monitors and integrates the conditions of all of its critical infrastructure, including roads, bridges, tunnels, rail/subways, airports, seaports, communications, water power, even major buildings, can better optimize its resources, plan its preventive maintenance activities, and monitor security aspects while maximizing services to its citizens"
Hollands (2008)	"Territories with a high capacity for learning and innovation, which is built on the creativity of their population, their institutions of knowledge production, and their digital infrastructure for communication"

(continued)

Table 1.1 (continued)

Authors	Definitions
Mattoni et al. (2015)	"An integrated and smart method of planning, assisted by the digital infrastructure for communication and management, would allow to coordinate the city as a sentient, homeostatic, and self-repairing organism. In other words, it could help this organism to behave as a resilient ecosystem. The regulation of the whole system through a dynamic balance is stimulated by the knowledge of the interrelations among subsystems and the real-time management of transformations"
Paskaleva (2009)	"A city that takes advantage of the opportunities offered by ICT in increasing local prosperity and competitiveness—an approach that implies integrated urban development involving multi-actor, multi-sector, and multi-level perspectives"
Piro et al. (2014)	"A city that intends as an urban environment which, supported by pervasive ICT systems, is able to offer advanced and innovative services to citizens in order to improve the overall quality of their lives"
Yigitcanlar et al. (2019)	"An ideal model to build the cities of the 21st century, in the case, its practice involves a system of systems approach and a sustainable and balanced view on the economic, societal, environmental, and institutional development domains"
Washburn et al. (2010)	"Smart city is a city that uses smart computing technologies to make critical infrastructure components and services more efficient, intelligent, and interconnected"

Source: own work based on the cited literature and Yigitcanlar et al. (2019)

civilisational achievements of the twentieth and twenty-first centuries (Caragliu et al., 2011; Bakici et al., 2013; Odendaal, 2003; Paskaleva, 2009), whereby, these technologies are to be used in urban infrastructure and to serve the needs of the local community. In the cited definitions, the need to integrate all urban technical systems and to monitor and control their operations also draws attention. In essence, the use of the latest technologies as well as their integration should be oriented towards improving the efficiency and effectiveness of their use, which is also a source of advantage of smart cities over "ordinary" cities (Janik et al., 2020; Kitchin, 2014; Lazaroiu & Roscia, 2012; Maalsen, 2019; Sokolov et al., 2019).

In light of the above, the definitional view of the Smart City is dominated by innovative and technological aspects. To some extent, but not in all definitions, social and environmental aspects appear. Arranging the definitions in chronological order, it can also be seen that the dominance of technological issues decreases over time and gives way to such concepts as quality of life, ecosystem, environment, or sustainability. This indicates a systematic evolution of the concept of Smart City in line with the expectations of theoreticians and practitioners associated with this concept. As a summary of the definitional review, Fig. 1.1 presents a schematic representation of the most important terminological determinants of Smart City.

In Fig. 1.1, among the outcomes of SC, actions, security has been added, which has not always been exposed in the definitions cited earlier. However, in connection with the COVID-19 pandemic, it has undoubtedly gained importance. The most probable reason for the omission of this aspect in previous definitions was the positive satisfaction of this need, since according to the public finance paradigm, urban security is a collective need felt and realised only in a situation of danger. At the same time, it is worth pointing out that the context of this safety should be multidimensional and refer not only to sanitary safety or public order, but also to threats connected with failures of urban systems, natural disasters, or accidents in industrial plants located in the city.

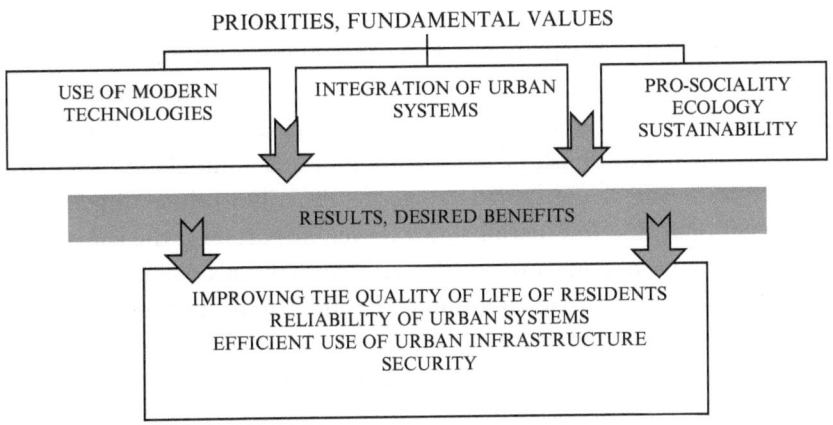

Fig. 1.1 Priorities and results of implementing the Smart City concept. Source: own work

The determinants of a Smart City in the literature are often described as separate thematic areas, which has led to the emergence and constitution of concepts such as (Lombardi et al., 2012):

- *Smart People* stand for high-quality human capital, characterised by specialised knowledge and skills achieved through advanced education,
- *Smart Governance* is associated with democratic, transparent, high-quality public management and involving all municipal stakeholders,
- *Smart Economy* is associated with the involvement of high technologies in production, services, and trade, so that they are a source of competitive advantage and efficiency,
- *Smart Mobility* is characterised by using IT, ICT, and high integration of transport systems and extensive use of clean mobility solutions,
- *Smart Environment* aimed at the sustainable use of resources and linked to a high level of environmental awareness in the urban community,
- *Smart Living* is associated with a high quality of life for residents and their multi-faceted safety.

In practice, the above dimensions of SC performance are the basis for evaluating and categorising cities in various rankings (Koca et al., 2021). The description characterising each of these dimensions is then parameterised by a set of quantifiable indicators. An example of such an evaluative set is presented in Table 1.2.

The information in Table 1.2 shows that most attention in the assessment of smart cities is paid to social aspects related to the quality of life (*smart living*) and human capital (*smart people*). In the case of mobility, which is the main subject of consideration in this monograph, the accessibility of means of transport is assessed primarily in three dimensions: local, national, and international. Moreover, important is the use of ICT systems in urban transport and logistics and the accessibility of these technologies to the urban community. In addition to the scope and level of development of transport infrastructure, qualitative logistics parameters such as sustainability, innovation, and safety of transport systems are also assessed.

The approach presented in the framework of the definition and evaluation of SC reflects the holistic nature of the concept analysed, which

Table 1.2 Areas, factors, and indicators for assessing smart cities in the context of their characteristics

Area	Evaluation factors and indicators
Smart economy	Innovative spirit (R&D expenditure in % of GDP; employment rate in knowledge-intensive sectors; patent applications per inhabitant)
	Entrepreneurship (self-employment rate; new businesses registered)
	Economic image & trademarks (importance as decision-making centre)
	Productivity (GDP per employed person)
	Flexibility of the labour market (unemployment rate; proportion of part-time employment)
	International embeddedness (companies with HQ in the city quoted on the national stock market; air transport of passengers; air transport of freight)
Smart people	Level of qualification (importance as knowledge centre; population qualified at levels 5–6 ISCED; foreign language skills)
	Affinity to lifelong learning (book loans per resident; participation in lifelong learning in %; participation in life-long learning in %)
	Social and ethnic plurality (share of foreigners; share of nationals born abroad)
	Flexibility (perception of getting a new job)
	Creativity (share of people working in creative industries)
	Cosmopolitanism (voter turnout at European elections)
	Open-mindedness (immigration-friendly environment (attitude towards immigration); knowledge about the EU)
	Participation in public life (voter turnout at city elections; participation in voluntary work)
Smart governance	Participation in decision-making (city representatives per resident; political activity of inhabitants; importance of politics for inhabitants; share of female city representatives)
	Public and social services (expenditure of municipal per resident in PPS; share of children in day care; satisfaction with quality of schools)
	Transparent governance (satisfaction with transparency of bureaucracy; satisfaction with fight against corruption)

(continued)

Table 1.2 (continued)

Area	Evaluation factors and indicators
Smart mobility	Local accessibility (public transport network per inhabitant; satisfaction with access to public transport; satisfaction with quality of public transport)
	(Inter-)national accessibility (international accessibility)
	Availability of ICT-infrastructure (computers in households; broadband internet access in households)
	Sustainable, innovative, and safe transport systems (green mobility share (non-motorised individual traffic); traffic safety; use of economical cars)
Smart environment	Attractivity of natural conditions (sunshine hours; green space share)
	Pollutions (summer smog (ozone); particulate matter; fatal chronic lower respiratory diseases per inhabitant)
	Environmental protection (individual efforts on protecting nature; opinion on nature protection)
	Sustainable resource management (efficient use of water (use per GDP); efficient use of electricity (use per GDP))
Smart living	Cultural facilities (cinema attendance per inhabitant; museum visits per inhabitant; theatre attendance per inhabitant)
	Health conditions (life expectancy; hospital beds per inhabitant; doctors per inhabitant; satisfaction with the quality of the health system)
	Individual safety (crime rate; death rate by assault; satisfaction with personal safety)
	Housing quality (share of housing fulfilling minimal standards; average living area per inhabitant; satisfaction with personal housing situation)
	Education facilities (students per inhabitant; satisfaction with access to the educational system; satisfaction with quality of the educational system)
	Touristic attractivity (importance as tourist location (overnights, sights); overnights per year per resident)
	Social cohesion (perception of personal risk of poverty; poverty rate)

Source: Giffinger et al. (2007a, b)

includes technological, economic, social, and environmental aspects of city life. It also allows us to see the multiplicity and complexity of the criteria that cities need to improve to be called smart places to live.

The above considerations show that the Smart City concept is based on very noble assumptions and is oriented towards the comprehensive development of local and regional communities. In the opinion of the authors of this monograph, the pursuit of excellence considered in the context of the quality of life of the inhabitants is, by all means, useful and desirable. Nevertheless, this concept is also subject to criticism due to the following circumstances (Bina et al., 2020; Kummitha, 2020; Colding & Barthel, 2017):

- utopianism, manifested in the assumption that social and economic inequalities can be eliminated and all SC residents can live better lives,
- exclusivity, due to the capital-intensive and high cost of living in SC,
- fostering the widening of social gaps between residents of "normal" cities and villages and those living in SC,
- excessive concentration and exposure to the convenience of life are closely linked to material status and participation in the use of goods and services considered luxuries,
- orienting the concept towards young and healthy people, while ignoring seniors or people with disabilities,
- widening civilisational, social, and economic differences between different regions of the world (Europe, with the largest number of smart cities, vs. Africa, with virtually no smart cities).

A response to criticism of the Smart City concept, including the objections described above, is its systematic and evolutionary development described in more detail in the next section. Its essence is to modify and complement the assumptions of creating smart cities aimed at eliminating the flaws and imperfections of the original SC concept.

The Evolution of the Smart City: From the High-Tech City to the Sustainable City

As already signalled in the previous section, the concept of the Smart City has systematically evolved over the last 40 years and will certainly continue to change in the future (Zheng et al., 2020; Lim et al., 2019; Bibri

& Krogstie, 2017). This is the result of technological, socio-economic, and civilisational changes that affect the scope and scale of individual and collective needs. People's expectations of quality of life are therefore changing, and this is reflected in the ultimate holistic dimension of smart cities.

The origin of the Smart City concept dates to the 1980s and is associated with the intensive development and use of IT and ICT technologies. The city, due to the concentration of population and economic life, on the one hand, became a place that triggered innovation, and on the other a testing ground for the mass implementation of all kinds of IT and telecommunications solutions. However, this perception of the city led to an unbalanced relationship between technology providers (IT and ICT)—city authorities and the local community. In this relationship, the business party located commercial solutions within the city, treating it as a traditional marketplace to the exclusion of identifying local and regional demands (Šurdonja et al., 2020; Kummitha, 2019). This is how the Smart City of generation 1.0 (Jucevičius et al., 2014) came into being and functioned, in which modern technologies and thus mainly the technological and economic dimensions of urban life matter (Barba-Sánchez et al., 2019).

Recovering the necessary balance between key stakeholders was restored by using the concept of the quadruple helix, according to which city authorities work closely with the scientific sphere and the local community to identify local needs and expectations, and only later make a concrete demand for IT or ICT products and services (Chen et al., 2017; McNeill, 2015). Additionally, on their own or in cooperation with representatives of science, they join the process of creating Smart City solutions becoming their driving force and co-creators. It is then possible for city authorities to consciously participate in creating and controlling the development of smart cities necessary to maintain a balance between the commercial, public, and social spheres. This is how generation 2.0 smart cities are emerging (Kumar et al., 2019; Hui et al., 2017). In these cities, electric transport, internet access, the Internet of Things, and the use of big data for urban management are rapidly developing (Yigitcanlar et al., 2019; Moser, 2015; Castells, 1989, 1996). Nevertheless, even at this stage of development, the main and active stakeholders of smart cities remain city authorities, business representatives, and academia. Although the needs of the local community are taken into account, the community itself does not participate in decisions and does not define the main areas of development. Its role and scope of influence changes only in the next generation of the Smart City, referred to as 3.0. Such a city unit is characterised by the active attitude of its inhabitants, who are willing to join the process of city management and want to consciously participate in it. In addition, in a Smart City of generation 3.0, another fivefold economic helix is realised (Leydesdorff, 2020). Science, city authorities, business, and the community are joined by environmental organisations that care for the urban ecosystem and minimise the city's adverse impact on the environment. In such an arrangement, it is possible to act according to the idea of sustainability, which means that each stakeholder satisfies its own needs and pursues its own priorities (Hatuka & Zur, 2020; Kummitha & Crutzen, 2019; Kummitha, 2019), while not compromising the well-being of the others (De Guimarães et al., 2020; Sakiewicz et al., 2020; Mattoni et al., 2019; Patel & Doshi, 2019; Kobza & Hermanowicz, 2018). This is also how a city that is not only smart but also sustainable emerges and operates (Ahvenniemi et al., 2017).

The idea of Smart City 3.0 is difficult to implement in practice due to idealistic assumptions about full participation and equal satisfaction of needs. In reality, the participation of the community in the decision-making process may turn out to be negligible and the influence of environmental organisations on the activities of city authorities may turn out to be completely insignificant. This in turn may result both from the inactivity of the stakeholders mentioned and from their much weaker negotiating position, requiring the support and openness of the municipal authorities.

When analysing the development of smart cities around the world, it can be seen that most of them are located in the most developed regions of the world, mainly in Europe. In Europe, on the other hand, the highest number and level of development of smart cities is characteristic of economies that are highly developed in terms of civilisation and economy. As a result, the leading positions in various rankings are almost always occupied by cities such as Copenhagen or Helsinki, representing the Scandinavian social economy with a high level of understanding for the idea of participation and a high level of economic, civilisational, and environmental development. These conditions are certainly conducive not only to the creation of smart cities, but also to their evolution towards a Smart City of generation 3.0.

In Poland, where the research conducted in this monograph is located, there are about 930 cities. The results of empirical analyses so far show that only a few of them can aspire to be smart in full (Jonek-Kowalska & Wolniak, 2021b; Kramarz et al., 2021). Nevertheless, the idea of SC is widely known and accepted, and cities would like to apply for Smart City status. A key barrier to this is the difficult economic and financial conditions (Jonek-Kowalska & Wolniak, 2021a). The average level of economic growth, the traditional profile of the economy, and the inadequacy of public funds as a source of financing for cities mean that expensive Smart City infrastructure solutions cannot be implemented and used extensively. Nevertheless, many cities and even villages try to improve the quality of life of their inhabitants through small modifications and improvements in the smart area, using cooperation with business and innovations of a less costly, organisational, or process nature.

As a result of the different perspectives on the Smart City and the evolution of the concept, the literature identifies four approaches to its analysis (Kummitha & Crutzen, 2017):

1. *restrictive school of thought,* which is based on the original definition of SC, focused on the use of modern IT and ICT technologies, in this view Smart City solutions are considered as tools for improving connectivity and collection, processing, and use of data relating to city functioning,
2. *reflective school of thought,* which assumes that the city serves not only to exploit modern technologies, but can also co-create them by using the potential of the local community, including above all its innovation and entrepreneurship,
3. *rationalistic school of thought,* which emphasises the role of human capital in creating a Smart City, and reduces IT and ICT technologies to the role of tools serving the development of this capital and the improvement of the quality of life in a Smart City,

4. *critical school of thought*, which criticises the concept of smart cities for, among other things: the appropriation and privatisation of urban space, the unethical use of extracted data, the generation and reinforcement of various types of exclusion and the limiting role of human capital (Basu, 2019; Bunders & Varró, 2019; Grossi & Pianezzi, 2017).

The approaches described above are also reflected in the way smart cities are described and studied, which either focuses on *technology-driven methods* (TDM) or refers mainly to *human-driven methods* (HDM). The first of the mentioned methodologies is most often used in publications concerning issues in the area of science or engineering. The second indicated methodology is used in social and humanistic analyses (Batabyal & Beladi, 2019; Camero & Alba, 2019; Lu et al., 2019).

Such a polarisation of methodology and concentration on selective aspects of Smart City operation is undoubtedly an advantage of specialised detailed considerations, in which what matters is the solution of a concrete technological or social problem. It should be remembered, however, that, as it was emphasised in the introduction, the city is an interdisciplinary, multidimensional, and dynamic system, which enforces the necessity to study and analyse the impact of individual solutions on its overall functioning. Social, humanistic, or environmental aspects must be balanced with technical and technological issues. Otherwise, the idea of a Smart City will be distorted and unsustainable.

Due to the above circumstances, in the remainder of this monograph the authors use research tools from both methodological groups. In the case of *human-driven methods*, they use instruments from the area of social sciences (e.g. surveys). In the case of *technology-driven methods*, they refer to modern, intelligent logistics solutions applied in urban infrastructure. Urban logistics is an excellent example of an area of SC activity that cannot be studied and analysed only with the use of one of the presented methodologies, because logistics systems are closely related to individual needs for the movement of people and materials, products, and goods.

Moreover, for similar reasons, the considerations and research carried out are embedded in the *rationalistic school of thought*, treating it as the one that best reflects the essence of the functioning of a Smart City, in which modern technologies are tools to serve the community and improve the quality of life of residents.

The authors also recognise the disadvantages of the Smart City as highlighted by the *critical school of thought*, but they treat them as certain barriers to the creation and functioning of smart cities, rather than as factors that should lead to the abandonment of improving and creating smart cities. Several key circumstances contribute to this attitude. First of all, as highlighted in the first section, all regions of the world are experiencing increasing urbanisation. It seems, therefore, that this process, due to its dynamics and mass character, can no longer be stopped. Cities have become centres of social and economic life, which are constantly expanding and developing. They are also important generators of income for local, regional, and national budgets, which means that the authorities at each of these levels have a real interest in their further development.

In the context of the conditions mentioned above, the only solution is to focus on improving the quality of life in cities, so that existing and emerging cities can become not only smart but also fully sustainable. Such an approach will make it possible, on the one hand, to maximise the effects of the use of IT and ICT solutions and, on the other hand, to eliminate the shortcomings of SC related to exclusion, lack of respect for the environment, or the commercialisation of urban space.

References

Ahvenniemi, H., Huovila, A., Pinto-Seppä, I., & Airaksinen, M. (2017). What are the differences between sustainable and smart cities? *Cities, 60*(Part A), 234–245. https://doi.org/10.1016/j.cities.2016.09.009

Angelidou, M. (2014). Smart city policies: A spatial approach. *Cities, 41*, 3–11. https://doi.org/10.1016/j.cities.2014.06.007

Bakici, T., Almirall, E., & Wareham, J. (2013). A smart city initiative: The case of Barcelona. *Journal of the Knowledge Economy, 4*(2), 135–148. https://doi.org/10.1007/s13132-012-0084-9

Barba-Sánchez, V., Arias-Antúnez, E., & Orozco-Barbosa, L. (2019). Smart cities as a source for entrepreneurial opportunities: Evidence for Spain. *Technological Forecasting and Social Change, 148*, 119713. https://doi.org/10.1016/j.techfore.2019.119713

Barrionuevo, J. M., Berrone, P., & Ricart, J. E. (2012). Smart cities, sustainable progress. Opportunities for urban development. *IESE Insight, 14*, 50–57.

Basu, I. (2019). Elite discourse coalitions and the governance of 'smart spaces': Politics, power and privilege in India's smart cities mission. *Political Geography, 68*, 77–85. https://doi.org/10.1016/j.polgeo.2018.11.002

Batabyal, A. A., & Beladi, H. (2019). The optimal provision of information and communication technologies in smart cities. *Technological Forecasting and Social Change, 147*, 216–220. https://doi.org/10.1016/j.techfore. 2019.07.013

Bibri, S. E., & Krogstie, J. (2017). Smart sustainable cities of the future: An extensive interdisciplinary literature review. *Sustainable Cities and Society, 31*, 183–212. https://doi.org/10.1016/j.scs.2017.02.016

Bina, O., Inch, A., & Pereira, L. (2020). Beyond techno-utopia and its discontents: On the role of utopianism and speculative fiction in shaping alternatives to the smart city imaginary. *Futures, 115*, 102475. https://doi. org/10.1016/j.futures.2019.102475

Bowerman, B., Braverman, J., Taylor, J., Todosow, H., & Wimmersperg, U. (2000). The vision of a Smart City, Paris. In *2nd International life extension technology workshop*.

Brol, R., Maj, M., & Strahl, D. (1990). *Metody typologii miast*. Wydawnictwo Akademii Ekonomicznej we Wrocławiu.

Bunders, D. J., & Varró, K. (2019). Problematizing data-driven urban practices: Insights from five Dutch 'smart cities'. *Cities, 93*, 145–152. https://doi. org/10.1016/j.cities.2019.05.004

Camero, A., & Alba, E. (2019). Smart City and information technology: A review. *Cities, 93*, 84–94. https://doi.org/10.1016/j.cities.2019.04.014

Caragliu, A., Del Bo, C., & Nijkamp, P. (2011). Smart cities in Europe. *Journal of Urban Technology, 18*(2), 65–82. https://doi.org/10.1080/1063073 2.2011.601117

Castells, M. (1989). *The Informational City: Information technology, economic restructuring, and the urban-regional process*. Blackwell.

Castells, M. (1996). *The information age: Economy, society and culture* (The rise of the network society) (Vol. 1). Blackwell.

Chen, T. M. (2010). Smart Grids, Smart Cities need better networks. *IEEE Network, 24*(2), 2–3.

Chen, Y., Ardila-Gomez, A., & Frame, G. (2017). Achieving energy savings by intelligent transportation systems investments in the context of smart cities. *Transportation Research Part D: Transport and Environment, 54*, 381–396. https://doi.org/10.1016/j.trd.2017.06.008

Colding, J., & Barthel, S. (2017). An urban ecology critique on the "Smart City" model. *Journal of Cleaner Production, 164*, 95–101. https://doi.org/10.1016/j.jclepro.2017.06.191

Czornik, M. (2004). Miasto. Ekonomiczne aspekty funkcjonowania, Prace Naukowe Akademii Ekonomicznej w Katowicach, Katowice.

De Guimarães, J. C. F., Andréa Severo, E., Felix Júnior, L. A., Leite Batista Da Costa, W. P., & Salmoria, F. T. (2020). Governance and quality of life in smart cities: Towards sustainable development goals. *Journal of Cleaner Production, 25320*, 119926. https://doi.org/10.1016/j.jclepro.2019.119926

Giffinger, R., Fertner, C., Kramar, H., Kalasek, R., Pichler-Milanović, N., & Meijers, E. (2007a). *Smart cities: Ranking of European medium-sized cities.* Vienna University of Technology.

Giffinger, R., Fertner, C., Kramar, H., Kalasek, R., Pichler-Milanovic, N., & Meijers, E. (2007b). *Smart cities: Ranking of European medium-sized cities* (pp. 5–24) Accessed 17 July 2021 http://www.smartcities.eu/download/smart_cities_final_report.pdf

Grossi, G., & Pianezzi, D. (2017). Smart cities: Utopia or neoliberal ideology? *Cities, 69*, 79–85. https://doi.org/10.1016/j.cities.2017.07.012

Hall, R. E., (2000). The vision of a Smart City. In *2nd International life extension technology workshop* (pp. 1–6).

Hatuka, T., & Zur, H. (2020). From smart cities to smart social urbanism: A framework for shaping the socio-technological ecosystems in cities. *Telematics and Informatics.*, In press, Available online 20 June 2020, Article 101430. https://doi.org/10.1016/j.tele.2020.101430

Hollands, R. G. (2008). Will the real smart city please stand up? *City: Analysis of Urban Trends Culture, Theory, Policy, Action, 12*(3), 303–320. https://doi.org/10.1080/13604810802479126

Hui, T. K. L., Sherratt, R. S., & Sánchez, D. D. (2017). Major requirements for building smart homes in smart cities based on internet of things technologies. *Future Generation Computer Systems, 76*, 358–369.

Janik, A., Ryszko, A., & Szafraniec, M. (2020). Scientific landscape of smart and sustainable cities literature: A bibliometric analysis. *Sustainability, 12*(3), 779. https://doi.org/10.3390/su12030779

Jonek-Kowalska, I., & Wolniak, R. (2021a). Economic opportunities for creating smart cities in Poland. Does wealth matter? *Cities, 114*, 103222.

Jonek-Kowalska, I., & Wolniak, R. (2021b). The influence of local economic conditions on start-ups and local open innovation system. *Journal of Open Innovation: Technology, Market, and Complexity, 7*(2), 110. https://doi.org/10.3390/joitmc7020110

Jucevičius, R., Patašienė, I., & Patašius, M. (2014). Digital dimension of Smart City: Critical analysis. *Procedia - Social and Behavioral Sciences, 15626*, 146–150. https://doi.org/10.1016/j.sbspro.2014.11.137

Kitchin, R. (2014). The real-time city? Big data and smart urbanism. *GeoJournal, 79*(1), 1–14. https://doi.org/10.1007/s10708-013-9516-8

Kobza, N., & Hermanowicz, M. (2018). How to use technology in the service of mankind? Sustainable development in the city. *IFAC-Papers, 51*, 340–345. https://doi.org/10.1016/j.ifacol.2018.11.328

Koca, G., Egilmez, O., & Akcakaya, O. (2021). Evaluation of the smart city: Applying the dematel technique. *Telematics and Informatics, 62*, 101625.

Kramarz, M., Przybylska, E., & Wolny, M. (2021). Reliability of the intermodal transport network under disrupted conditions in the rail freight transport. *Research in Transportation Business & Management*, 100686. https://doi.org/10.1016/j.rtbm.2021.100686

Kumar, H., Singh, M. K., & Gupta, M. P. (2019). A policy framework for city eligibility analysis: TISM and fuzzy MICMAC-weighted approach to select a city for smart city transformation in India. *Land Use Policy, 82*, 375–390. https://doi.org/10.1016/j.landusepol.2018.12.025

Kummitha, R. R. K. (2019). Smart cities and entrepreneurship: An agenda for future research. *Technological Forecasting and Social Change, 149*, 119763. https://doi.org/10.1016/j.techfore.2019.119763

Kummitha, R. K. R. (2020). Why distance matters: The relatedness between technology development and its appropriation in smart cities. *Technological Forecasting and Social Change, 157*, Article 120087. https://doi.org/10.1016/j.techfore.2020.120087

Kummitha, R. K. R., & Crutzen, N. (2017, July). How do we understand smart cities? *An Evolutionary Perspective Cities, 67*, 43–52. https://doi.org/10.1016/j.cities.2017.04.010

Kummitha, R. K. R., & Crutzen, N. (2019). Smart cities and the citizen-driven internet of things: A qualitative inquiry into an emerging smart city. *Technological Forecasting and Social Change, 140*, 44–53. https://doi.org/10.1016/j.techfore.2018.12.001

Lazaroiu, G. C., & Roscia, M. (2012). Definition methodology for the smart cities model. *Energy, 47*(1), 326–332. https://doi.org/10.1016/j.energy.2012.09.028

Leydesdorff, L. (2020). The triple helix: An evolutionary model of innovations. *Research Policy, 29*(2), 243–255. https://doi.org/10.1016/S0048-7333(99)00063-3

Lim, Y., Edelenbos, J., & Gianoli, A. (2019). Identifying the results of smart city development: Findings from systematic literature review. *Cities, 95*, 102397. https://doi.org/10.1016/j.cities.2019.102397

Lom, M., & Pribyl, O. (2020). Smart city model based on systems theory. *International Journal of Information Management*. In press, available online 19 February 2020, article no. 102092. https://doi.org/10.1016/j.ijinfomgt.2020.102092

Lombardi, P., Giordano, S., Farouh, H., & Yousef, W. (2012). Modelling the Smart City performance. *Innovation: The European Journal of Social Science Research, 25*(2), 137–149.

Lu, H.-P., Chen, C.-S., & Yu, H. (2019). Technology roadmap for building a smart city: An exploring study on methodology. *Future Generation Computer Systems, 97*, 727–742. https://doi.org/10.1016/j.future.2019.03.014

Maalsen, S. (2019). Smart housing: The political and market responses of the intersections between housing, new sharing economies and smart cities. *Cities, 84*, 1–7.

Malarski, S. (2000). Status prawno-administracyjny miasta w ustawodawstwie II, -i-III Rzeczypospolitej. W: Społeczne, gospodarcze i przestrzenne przeobrażenia miast, red. J. Słodczyk, Wydawnictwo Uniwersytetu Opolskiego, Opole.

Manczak, I. (2014). Ewolucja miasta. Od miasta tradycyjnego do miasta innowacyjnego. *Zeszyty Naukowe Uniwersytetu Ekonomicznego w Krakowie, 11*(935), 45–57.

Mattoni, B., Gugliermetti, F., & Bisegna, F. (2015). A multilevel method to assess and design the renovation and integration of Smart Cities. *Sustainable Cities and Society*, 105–119.

Mattoni, B., Nardecchia, F., & Bisegna, F. (2019). Towards the development of a smart district: The application of an holistic planning approach. *Sustainable Cities and Society, 48*, 101570. https://doi.org/10.1016/j.scs.2019.101570

McNeill, D. (2015). Global firms and smart technologies: IBM and the reduction of cities. *Transactions of the Institute of British Geographers, 40*(4), 562–574. https://doi.org/10.1111/tran.12098

Moser, S. (2015). New cities: Old wine in new bottles? *Dialogues in Human Geography, 5*(1), 31–35. https://doi.org/10.1177/2043820614565867

Odendaal, N. (2003). Information and communication technology and local governance: Understanding the difference between cities in developed and emerging economies. *Computers, Environment and Urban Systems, 27*(6), 585–607. https://doi.org/10.1016/S0198-9715(03)00016-4

Paskaleva, K. A. (2009). Enabling the smart city: The progress of city e-governance in Europe. *International Journal of Innovation and Regional Development, 1*, 405–422.

Patel, Y., & Doshi, N. (2019). Social implications of smart cities. *Procedia Computer Science, 155*, 692–697. https://doi.org/10.1016/j.procs.2019.08.099

Piro, G., Cianci, I., Grieco, L. A., Boggia, G., & Camarda, P. (2014). Information centric services in smart cities. *Journal of Systems and Software, 88*, 169–188.

Sakiewicz, P., Piotrowski, K., Ober, J. P., & Karwot, J. (2020). Innovative artificial neural network approach for integrated biogas - wastewater treatment system modelling: Effect of plant operating parameters on process intensification. *Renewable Sustainable Energy Review, 124*, 1–14. https://doi.org/10.1016/j.rser.2020.109784

Sokolov, A., Veselitskaya, A., Carabias, V., & Yildirim, O. (2019). Scenario-based identification of key factors for smart cities development. *Policies. Technological Forecasting & Social Change, 148*, 119729.

Sulemana, I., Nketiah-Amponsah, E., Codjoe, E. A., & Andoh, J. A. N. (2019). Urbanization and income inequality in sub-Saharan Africa. *Sustainable Cities and Society, 48*, 101544. https://doi.org/10.1016/j.scs.2019.101544

Šurdonja, S., Giuffrè, T., & Deluka-Tibljaš, A. (2020). Smart mobility solutions—necessary precondition for a well-functioning smart city. *Transportation Research Procedia, 45*, 604–611. https://doi.org/10.1016/j.trpro.2020.03.051

Washburn, D., Sindhu, U., Balaouras, S., Dines, R. A., Hayes, N. M., & Nelson, L. E., (2010). Helping CIOs understand "Smart City" initiatives: Defining the Smart City, its drivers, and the role of the CIO.

Yigitcanlar, T., Han, H., Kamruzzaman, M. D., Ioppolo, G., & Sabatini-Marques, J. (2019). The making of smart cities: Are Songdo, Masdar, Amsterdam, San Francisco and Brisbane the best we could build? *Land Use Policy, 88*, 104187. https://doi.org/10.1016/j.landusepol.2019.104187

Yigitcanlar, T., Kamruzzaman, M. D., Foth, M., Sabatini-Marques, J., da Costa, E., & Ioppolo, G. (2019). Can cities become smart without being sustainable? A systematic review of the literature. *Sustainable Cities and Society, 45*, 348–365.

Zhang, X., Wan, G., Luo, Z., & Wang, C. H. (2019). Explaining the East Asia miracle: The role of urbanization. *Economic Systems, 43*(2), 100697. https://doi.org/10.1016/j.ecosys.2019.100697

Zheng, C. H., Yuan, J., Zhu, L., Zhang, Y., & Shao, Q. (2020). From digital to sustainable: A scientometric review of smart city literature between 1990 and 2019. *Journal of Cleaner Production, 258*, 120689. https://doi.org/10.1016/j.jclepro.2020.120689

2

Areas of Logistical Support for Cities

Urban Logistics and City Logistics System

Considerations on the concept of urban logistics are directly related to the origins of the concept itself, which makes it possible to see that urban logistics emerged as a response to the needs of city residents, while at the same time becoming a kind of tool supporting the management of the city logistics system. Szołtysek (2010) emphasises that urban development and all flows (movements) within the city, both of freight and of people, occur independently of each other and lead to unnecessary strain on the linear transport infrastructure, increasing the intensity of competition between users for its capacity, because in large part these movements are not subject to coordination processes. These movements cause problematic effects, in particular transport congestion, an adverse impact on the city environment, a reduction in the level of customer service (freight receivers and shippers, individuals travelling within the city), and an increase in the cost of movement.

The certain inconvenience resulting from the increased intensity of flow, the deterioration of the city's functionality, and, consequently, the decline in residents' satisfaction with life has led many authors to define

M. Kramarz et al., *Urban Logistics in a Digital World*,
https://doi.org/10.1007/978-3-031-12891-2_2

the concept of urban logistics, which by offering certain logistical solutions was to become a kind of remedy for the existing undesirable complications. Rzeczyński (2007) even proposes to speak of: "logistics for (in the service of) the city" (Rzeczyński, 2007). The pioneers of research reaching for tools to optimise flows of cargo, people and information were the Japanese led by Taniguchi (2015), who states that urban logistics includes the process of optimising logistics activities and transport by private companies in urban areas (Kiba-Janiak & Witkowski, 2014). Bektaş et al. (2015) in turn captured the concept of urban logistics in three key aspects:

- in relation, to urban freight transport,
- based on the idea of an integrated logistics system with consolidation and coordination, and
- in the context of increasing productivity and reducing environmental damage.

Fontaine et al. (2021), on the other hand, state that in its most basic sense, urban logistics seeks to reduce the nuisance associated with the transport of goods within urban areas, while sustaining the social and economic development of the organisations and cities involved (Bektas et al., 2017; Crainic et al., 2020; Savelsbergh & Van Woensel, 2016; Taniguchi et al., 2001). Much of the urban logistics literature deals with the development of optimisation models and algorithms to solve strategic (e.g., Gianessi et al., 2016), tactical (e.g., Crainic et al., 2009) or operational distribution planning problems (e.g. Barceló et al., 2007).

Urban or City Logistics?

The terminology used in the field logistics in the city can be very controversial. Many researchers use different nomenclature when attempting to define a concept and based on a given concept or scope. Tundys (2008a, b) helps to distinguish between the two concepts by explaining that city logistics encompasses the whole system, while urban logistics encompasses the individual subsystems in the system and deals with the implementation of specific concepts in the subsystems. Urban logistics, in contrast to city logistics, is therefore characterised by a more holistic view of the city area and the flow of information, people, and goods within it. Szołtysek (2010), on the other hand, recommends treating both concepts as synonyms, explaining that it does not make much sense to divide the application of logistics for city needs into fragmentary and comprehensive. The unnaturalness of the division is also shown by Szymczak (2006), who points out that it is not possible to consider city centres without other areas where transport/logistical events also occur. Relieving congestion in the centre does not make the whole city unobstructed, but may cause congestion in other parts of the city, above all near the centre (Szymczak, 2006). Tundys (2013) emphasises that different nomenclature can contribute to blurring the definitional "purity", and the multiplicity of terms introduces chaos in presenting the scope of logistics impact in the city. The author considers that the concept of urban logistics will therefore suffice and proposes the following general definition: **urban logistics** is the links of the logistics system within a defined urban area, with particular emphasis on the transport system covering flows of both goods and people, aiming at efficient, optimal, and ecological coordination of all movements in the city (and in special cases extending beyond its administrative boundaries—e.g. passenger transport to/from neighbouring municipalities). It aims to efficiently, optimally, and ecologically coordinate all movements in a city (and in specific cases beyond its administrative borders—e.g. passenger transport to/from neighbouring municipalities), covering all kinds of components: infrastructural, organisational, information-technological, and human, taking into account economic, social, and environmental aspects. Concepts supporting the functioning of the urban logistics system, relieving the environment, urban traffic, and its intra-urban space, being an element of logistical support of urban structures, are considered as projects in the field of urban logistics (Tundys, 2013). Urban logistics can also be presented in a system perspective as a set of elements and the occurring relationships and relations between its components. Taking this approach into account, an **urban logistics system** can be presented as a purposefully organised set of elements, including its stakeholders, infrastructure, regulatory norms and tariff systems, and the relationships between them, which are involved in the flow of people, cargo, and accompanying information in urbanised areas (Kiba-Janiak & Witkowski, 2014). Based on the definition of the urban logistics system, it is possible to distinguish elements of a tangible nature: transport, superstructure, enterprises, cargo and technical infrastructure of transport and municipal economy and of an intangible nature, which are of an optional nature, resulting from procedures, standards or good practices, or of an obligatory nature, resulting from the need to comply with the law: information flows, regulatory norms, tariff system (Kiba-Janiak & Witkowski, 2014).

Depending on the extent of logistics influence on a given area, urban logistics can be divided in the context of several levels (Fig. 2.1) into **macro (Macro City Logistics)**, where the research object is the city as a holistic level and general economic relations within the city; **micro (Micro City Logistics)**, where the research object is a single economic relation and links between them in a given area, an important aspect here being the process of distribution of these objects (e.g. to households or economic institutions) and methods of their implementation; **meta (Meta City Logistics),** where the objects of research are horizontal

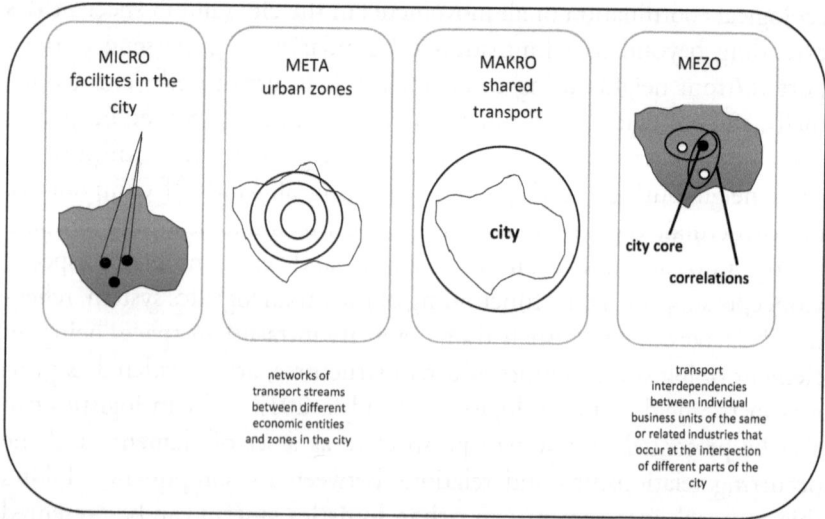

Fig. 2.1 Spatial relationships between the planes of the city logistics system. Source: Tundys (2013)

economic relations between economic entities located inside the urban space; and **mezo (Mezo City Logistics)**, where the object of research are districts and specific areas of the city, there is a division into zones of influence and supply, as well as connections between city zones often of the same or related industries (Tundys, 2013).

To meet the social and economic needs of the users of urban space and thus improve the quality of life of citizens, urban logistics must have a well-defined set of goals and objectives. Among them, we can distinguish the economic goal related to the costs of city operation and the most possible reduction of them, the environmental goal oriented towards reducing the negative effects of urban logistics, taking into account respect for the environment, and the social goal, oriented towards the inhabitants, the degree of their satisfaction and improvement of the quality of life in the city (Tundys, 2013). Realising the mentioned triad of main objectives of urban logistics, understanding the relationships that occur between the given areas, requires an interdisciplinary approach. Therefore, it is helpful to draw on the research achievements of other scientific disciplines, for example, ecology, transport policy, urban planning, or social marketing. All this contributes to the integrative character of logistics management.

City Logistics System

The city has a number of characteristics that allow to define its structure as systemic. The literature on the subject gives many approaches to the classification of the logistics system of the city, resulting primarily from the functions it is to perform, while the appropriate configuration of the system elements allows you to shape its parameters in such a way that it can function properly to meet the needs of users. Kiba-Janiak (2018a, b) proposes a division according to the criterion:

- subject matter, focusing on flows of people, freight, and information,
- functional relating to the subsystem of transport, storage, waste management, customer service, and location activities.

In turn, Iwan (2013a, b) classifies the city logistics system into five basic subsystems:

- transport of material goods,
- transport of passengers,
- transport and storage of waste,
- storage of material goods,
- control of flow of material, goods, and persons (integrating the subsystems mentioned).

Szołtysek (2016) includes the following elements in the breakdown of the city logistics system:

- transport and storage of material goods and municipal waste,
- public transport,
- individual transport,
- pedestrians, cycling, and other non-motorised travel subsystems,
- regulation subsystem—creating movements of material goods and people.

Taking into account the differences in the classification of city logistics subsystems, it can be concluded that it is necessary to adequately plan the course of logistics processes related to the movement of material goods and people. It should also be emphasised that all elements of the city logistics system are equifinal in relation to each other (Wiktorowska-Jasik, 2014). In addition, the city logistics system is strongly influenced by factors that are the result of diverse user needs related to mobility, manufacturing/production, learning and development, acquisition of goods, and recreation (Tundys, 2013).

Logistics is involved in building city attractiveness and economic growth, as well as improving quality of life (Holguin-Veras et al., 2021; Skultety et al., 2021). Urban logistics, as the research indicates, supports the urban economy and sustainability, but at the same time the negative externalities of freight and passenger transport increase as the demand for goods and services increases. Increasingly, cities are therefore taking the initiative to integrate urban logistics planning into urban planning

(Matusiewicz, 2019). In doing so, they are struggling to ensure a friendly environment around their road infrastructure, where the increasing presence of light and heavy trucks raises questions about safety and environmental impacts (Semanjski & Gautama, 2019). Among the main problems encountered in urban logistics in terms of freight flows are (Szołtysek, 2009a, b, c; Ros-McDonnell et al., 2018):

- access to line infrastructure for delivery vehicles, mainly due to insufficient infrastructure, limited access, or congestion,
- environmental aspects: harmful emissions, noise, vibrations, and physical obstructions,
- safety: heavy goods vehicles, due to their size, manoeuvrability, and loading/unloading operations on streets, often cause accidents,
- energy consumption: urban transport is the most fuel-consuming sector.

When analysing logistics problems in a city and the decisions made to solve them, decision-makers are confronted with often different expectations of stakeholders. At this point, the question arises: How to evaluate decisions and logistical solutions in a city?

The evaluation aspect of urban logistics is covered very differently in the literature. Shafique et al. (2020) demonstrated the link between freight transport, economic prosperity, urbanisation, and $CO2$ in cities. Bosona (2020) focuses on the evaluation of urban decision-making in relation to the growth of e-commerce, which has significantly contributed to the growth of urban freight traffic, both in terms of freight mass and freight movement. The assessment is related to the extent to which the objectives related to sustainable urban development are met. Similar studies are also conducted by other authors. For example, research on the impact and challenges of new technologies in urban mobility was conducted by Paiva et al. (2001). Among the key trends also influencing urban logistics, and with it the quality of life of city residents, the authors identified: Internet of Things, artificial intelligence, blockchain, and big data technology point to their application in smart mobility for the development of the Smart City ecosystem. They define smart mobility as a concept that fits into the United Nation's 17 Sustainable Development Goals by 2030. The assessment conducted by these authors is therefore based on an indication of the degree of achievement of the adopted sustainable development goals.

Analysing the research on logistics solutions in the city, two variants of their evaluation can be noted:

- evaluation through the external costs of transport, thus achieving the objectives assigned to the sustainable development of the city,
- evaluation through impact on quality of life.

Of course, it is important to bear in mind the relationship between these approaches: an increase in external transport costs has a negative impact on quality of life. Thus, the final dimension of any urban solution (including logistics solutions) is the quality of life.

Quality of life, according to Taniguchi (who has presented this view in a number of publications including Taniguchi et al. (2001, 2003); Taniguchi & van der Heijden (2000)), is an important objective of urban logistics. There are many definitions of quality of life and each definition points to factors that are important from the author's point of view. Quality of life according to Kolman (2009) is "the degree of satisfaction of spiritual and material needs of human beings, the degree of satisfaction of requirements, determining the level of material and spiritual existence of individuals and society as a whole, the degree of fulfilment of expectations of conventional normality in activities and situations of daily life of individuals and society". Glazer (2012) considers that "quality of life is an individual's feeling of well-being, his satisfaction or dissatisfaction". Gillingham and Reece (1980) already gave a similar interpretation indicating that "quality of life is the level of satisfaction obtained by an individual as a result of the consumption of goods and services, leisure activities, and the enjoyment of other material and social conditions of the environment in which that individual finds himself". Liu (1975a, 1975b) had an interesting perspective on the interpretation of this concept, stating that "quality of life means a set of needs whose satisfaction makes people happy".

Zelias (2004), on the other hand, distinguishes between standard of living and quality of life. According to him, quality of life is most often described by qualitative characteristics, and standard of living by quantitative methods. A similar approach to quality of life was adopted by Borys and Rogala (2009), according to whom one can distinguish between objective and subjective quality of life. Objective quality of life is described using indicators such as, among others, monthly income or living space. Objective quality of life is similar (but not identical) in meaning to the concept of "living conditions" or "standard of living", which mean "the totality of objective conditions of an infrastructural nature in which a society lives". They are mainly connected with the material condition, existential security, and environmental security of an individual's life. Subjective quality of life, on the other hand, is an assessment of the degree to which needs are satisfied, for example, satisfaction with earned income, possibility to find a good job, safety, satisfaction with the flat one has, possibility to spend leisure time in an attractive way, efficient movement around the city, access to education, health care or convenient shopping, etc. According to Taniguchi et al. (2001), urban logistics enables the implementation of innovative solutions that can improve the quality of life in cities.

Quality of life is a very complex category with a difficult to define scope and interdisciplinary character. This category is formed by many levels. Research on various aspects of quality of life was conducted by Szołtysek and Otręba (2015). Based on a review of the literature, the authors identified over a dozen specific problems of the quality of life of the inhabitants (living conditions, material basis of existence, education, culture, physical culture, leisure time, mobility, neighbourhood, safety, health, acceptance, free services, public space), to which they assigned indicators. Quality of life covers many areas related to a person's everyday functioning. Among the areas affecting quality of life is the aspect of urban mobility. The needs of people to move around the city results from many of the quality of life considerations mentioned above. At the same time, many of these areas are significantly affected by passenger and/or freight transport.

Flow of People in the City

The functioning of cities is inextricably linked to enabling people to move, regardless of the reasons that trigger communication needs, as well as the way in which the space is covered (Szołtysek, 2005). Thus, the flows of people constitute one of the main streams in the logistics system of a city (next to the stream of matter, energy, and information) (Tundys, 2013) and the object of interest of logistics, which in relation to the city has long been associated mainly with freight transport (Kiba-Janiak & Cheba, 2011). The universality of transport needs is reflected in high mobility indices. They are understood as the average daily number of total journeys made by people per day, regardless of the form of transport (Wyszomirski, 2008). Transport needs result from the transport links that exist between individual urbanised, economic and residential centres. The exchange of people occurs within the framework of commuting to work, school, shopping, business trips, and recreational and leisure trips (Szołtysek, 2005). In other, more general terms, communication needs of professional, subsistence, recreational, and other nature are listed. Thus, it can be considered that communication needs do not have a homogeneous character (Wojewódzka-Król & Załoga, 2016). The aforementioned transport links depend on the size of the city (population), the specificity of the urban area, including the degree of its industrialisation, the characteristics of the inhabitants (including demographic structure, occupational structure and places of employment), and the spatial range of public transport (Wyszomirski, 2008).

The fulfilment of needs for the movement of persons may be carried out primarily by collective or individual transport. Collective transport means regular transport performed at the request of a local government transport organisers only within a single municipality, two or more municipalities, by agreement among the municipalities forming the communal interrelationship. Individual transport is opposed to collective transport. It is characterised by specific terms of communication and lack of regularity (Kiba-Janiak & Cheba, 2011). Collective transport is thus identified with public transport or urban transport, while individual transport is identified with private transport (Cichosz, 2014). In addition to these two forms of transport, Cichosz (2014) points to another, less

Table 2.1 Differences between individual and collective transport

Description	Collective transport	Individual transport
Source of funding	Dually: cities and users of this kind of transport	Users of this type of transport means
Regularity of transportation	High regularity of transportation	Lack or rare regularity of transportation
Transport conditions	A large number of people using one means of transport	A small number of people using the same means of transport
Costs	Low cost of movement for one passenger	High costs of movement
Privacy	Lack of privacy	High privacy
House-to-house movement (flexibility of movement)	Moving along designated and fixed routes	You can reach any place

Source: Kiba-Janiak and Cheba (2011)

popular form, which is group transport (taxis, carpooling). The main differences between the basic forms of transport (individual and collective) are shown in Table 2.1.

The previously indicated basic divisions of passenger transport in the city can be complemented by a classification indicating the following: surface transport (the most popular form of transport, e.g., bus, trolleybus, motorbike, urban railway), underground transport (underground railway), aboveground transport (monorail), and water transport (e.g. water tram) (Janczewski, 2020). The most common forms of transport, for example, buses, trolleybus, motorcycles, urban, railway, underground transport (metro), overground transport (monorail), and water transport (e.g. water tram) (Janczewski, 2020); transport within a city A, transport from another city B/C to city A, transport from city A to another city B/C, and transport from city B to city C passing through city A. These classifications are shown in Fig. 2.2.

The progressive development of cities, together with the enrichment of their inhabitants and the increasing standard of living, entails the tendency to develop individual, private, and car-based transport. This is a very problematic issue, mainly due to the increase in external costs and a number of other negative effects (Cavoli, 2021; Tomaszewska, 2015; Kiba-Janiak & Cheba, 2011). The most important of these are shown in Fig. 2.3.

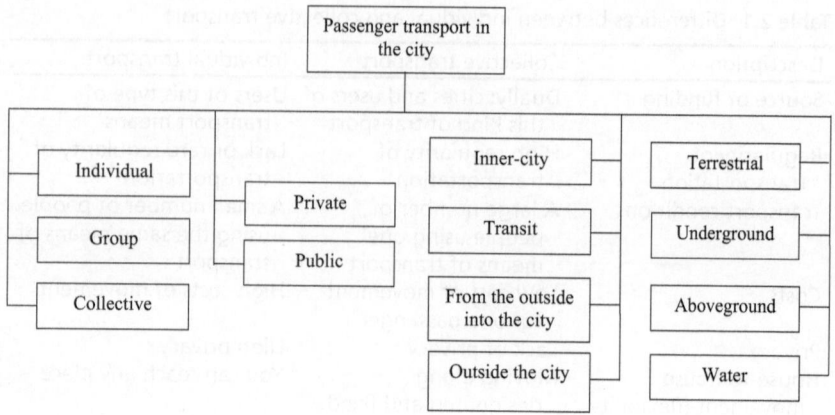

Fig. 2.2 Breakdown of passenger transport in the city. Source: own work based on: Cichosz (2014); Janczewski (2020)

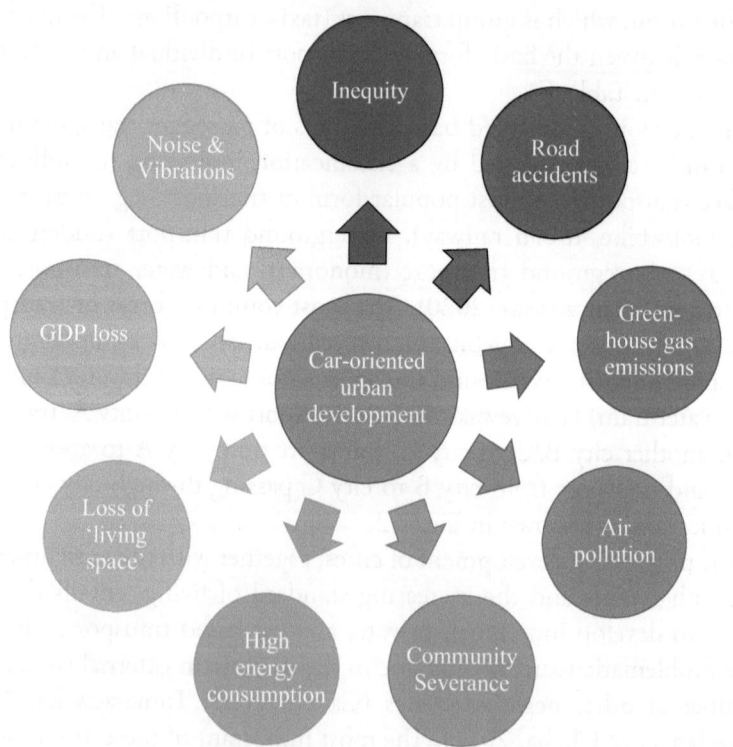

Fig. 2.3 Negative externalities associated with car-oriented urban development. Source: Cavoli (2021)

An alternative to individual transport is public collective transport, which is indicated as one of the solutions for the reduction of external transport costs in the city. It is considered as a sustainable form of transport, which apart from environmental benefits, compensates social disproportions connected with wealth and, consequently, with the ownership of means of transport (Krygsman et al., 2004). Public transport is an integral part of the urban culture of everyday life; it provides a basis for the interests of different social groups and thus meets the most important needs of society (Strychul et al., 2021). Above all, it has been recognised for its contribution to reducing greenhouse gas emissions, reducing traffic, improving air quality (Dillman et al., 2021; Jain & Tiwari, 2016; Kwan & Hashim, 2016), and mitigating climate change (Bulkeley & Castán Broto, 2013; Sethi et al., 2020). The challenge for public transport is the uptake of energy-efficient transport modes or low-emission vehicles. In addition, it needs to be connected to other forms of transport, including those in the area of micromobility, which will encourage people to shift away from individual transport (Wimbadi et al., 2021).

A major challenge for urban passenger transport is to convince the public to use public transport, especially as attempts to do so often fail. The basis for action is the knowledge of the requirements and expectations of potential transport users in relation to transport and the efforts of decision-makers to meet them (Krygsman et al., 2004). This is particularly important given that public transport services are often perceived as homogeneous, not very competitive, offered by operators, often operating in near-monopoly conditions and thus not adapted to individual user needs compared to other modes (bicycles or cars) (Mouwen, 2015). In relation to these problems, the role of local authorities, who are responsible for shaping transport and environmental policies in the city, is highlighted. The authorities should, in cooperation with other stakeholders (transport operators, residents, and public transport managers), develop a policy (strategy) for green logistics in the city, one element of which is to reduce individual transport in favour of public transport (Kiba-Janiak & Cheba, 2014).

When considering passenger transport, the requirements that are formulated for it cannot be ignored. These requirements are referred to as transport postulates. The most popular of them are: cost, time,

Table 2.2 Transport demand in urban passenger transport

Postulate	Interpretation
Directness	Connection without interchange
Frequency	Time intervals between departures of vehicles on the same line
Availability	Distance from the stop (spatial or temporal)
Reliability/assurance	Arrival at the destination at the appointed time
Low cost	Low, single, or periodic tariff charge
Speed	Driving time including stops en route
Punctuality	Compliance of departures with the timetable
Rhythmicity	Uniform time intervals between consecutive departures in the same direction
Comprehensive information	Methods of communicating the transport offer and changes to the offer
Convenience	The set of elements that determine waiting at a stop and boarding conditions

Source: Hebel and Wyszomirski (2016)

accessibility, and convenience (Zimon & Gosik, 2015; Hebel & Wyszomirski, 2016). A broader account of transport postulated in urban passenger transport is included in Table 2.2.

Fulfilment of the indicated postulates influences the level of quality of the provided transport activities, simultaneously influencing the level of satisfaction of transport users.

However, it is worth noting that these postulates have different importance for transport users. Allen et al. (2019) attempted to classify the parameters affecting the quality of passenger transport. For this purpose, they used the hierarchy included in Maslow's pyramid of needs, distinguishing three groups of attributes: functional (utilitarian), security (protection), and hedonic (excitement) (Fig. 2.4).

Passenger transport in the city coexists with freight transport within the same limited space. It is emphasised that freight transport and passenger transport compete for limited infrastructure resources. This leads to a number of irregularities and reduced efficiency of the operating economic entities, as well as to negative consequences for passenger transport participants (Tundys, 2013; Krysiuk & Nowacki, 2016). Current assumptions of urban transport management, both freight and passenger, are oriented towards the pursuit of sustainability, which is a major

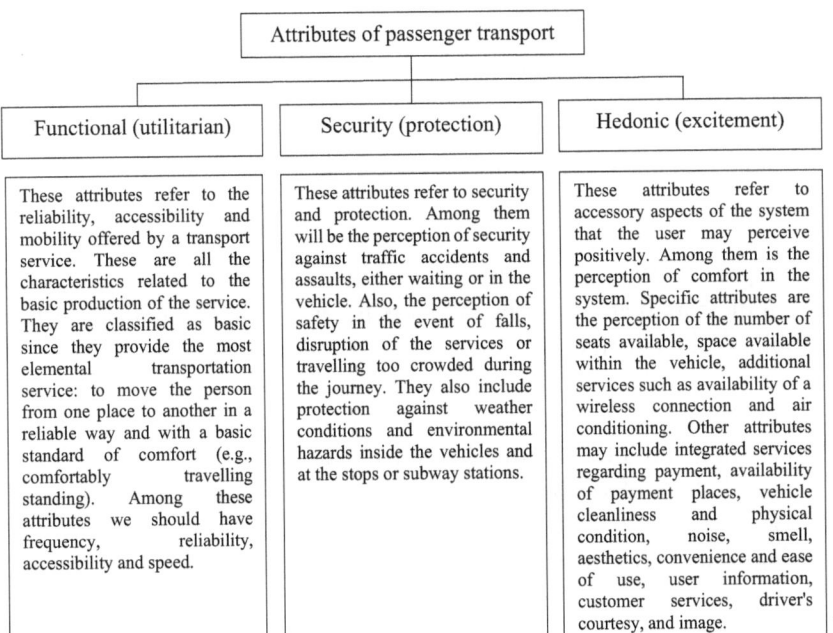

Fig. 2.4 Group of attributes for urban passenger transport. Source: compiled from Allen et al. (2019)

challenge worldwide (Zawieska & Pieriegud, 2018). Sustainable urban mobility has become an important global, regional, and local issue (Regmi, 2020). It should be emphasised that cities are particularly important areas for policies to reduce pollution and minimise other external costs of transport, as they concentrate most of the human population and their mobility activities. Moreover, the Smart City concept implies the pursuit of sustainable urban mobility (Zawieska & Pieriegud, 2018). A sustainable urban transport system is a movement system that allows meeting the transport-related needs of urban residents (Chamier-Gliszczyński, 2011):

- in a way that is safe, does not jeopardise human health or the environment, and maintains the balance between the generations,
- is affordable for all city residents,

- allows it to function effectively and sustains the economy and regional development,
- aims to reduce emissions of harmful gases,
- reduces the generation of waste,
- uses renewable energy sources,
- does not cause traffic congestion.

Regmi (2020), as part of his pilot study in four Asian cities, proposed ten indicators that can form the basis for measuring sustainable urban mobility. These include:

- the extent to which transport plans cover public transport, multimodal facilities, and infrastructure for active modes,
- modal share of active and public transport in commuting (per cent of trips),
- convenient access to public transport services (per cent of the population),
- public transport quality and reliability (per cent satisfied with service),
- traffic fatalities per 100,000 inhabitants (number of fatalities),
- affordability—travel costs as part of income (per cent of income),
- operational costs of the public transport system (cost recovery ratio),
- investment in public transport systems (per cent of total investment in transport),
- air quality in the city (PM 10) (ug m^3),
- greenhouse gas emissions from transport (tons/capita/year).

Designing sustainable flows of people in a city requires taking into account the problems of so-called social exclusion and aiming for high levels of social equity. Social exclusion can affect various social groups, in particular the elderly, the disabled, the poor, and those living in suburban areas (Sun & Lau, 2021; Yang, 2018; Rocha et al., 2021; Krygsman et al., 2004; Mogaji & Nguyen, 2021). Especially the progressive ageing of the population poses a major challenge in terms of mobility for older people, often with limited physical activity. Well-designed public transport can enable older people to maintain full mobility and independence after they stop driving themselves (Sun & Lau, 2021; Yang, 2018). Unfortunately, as Yang (2018) highlights, the problem of transport social exclusion is often ignored by government officials and urban transport decision-makers. The availability of well-developed public transport can also compensate for the disparities resulting from economic inequality between those with access to private transport and those of lower income without such access (Rudke et al., 2021; Krygsman et al., 2004). Adapting urban transport to the needs of people with disabilities is also a major challenge. Affordable, accessible, and reliable public transport is often the only form of travel to access the basic necessities of life (education, health care, employment, etc.) (Mogaji & Nguyen, 2021).

In the quest for urban passenger transport sustainability, the role of so-called shared mobility, based on the sharing of different types of transport modes, is emphasised. It is the result of significant developments in both technology and information and communication (Tirachini et al., 2020). This concept also includes micromobility, which is an innovative transport solution based on own and increasingly common shared lightweight devices (bikesharing, scootersharing, mopedsharing). Micromobility first and foremost makes it possible to replace the car over short distances and can thus be an advantageous solution for first- and last-mile transport. It contributes to changes in mobility patterns and the behaviour of transport participants. At the same time, it fits in well with multimodal travel and becomes part of an integrated transport system in a city (Abduljabbar et al., 2021; Oeschger et al., 2020).

Sustainable transport has become a major determinant of sustainable cities and at the same time one of the most important challenges of the twenty-first century. Nowadays, the construction of smart mobility is identified as a key factor contributing to the said sustainability. Benevolo et al. (2016) state that smart mobility "could be seen as a set of coordinated actions addressed at improving the efficiency, effectiveness and environmental sustainability of cities". According to Debnath et al. (2014), to consider a city's mobility system as smart, it is necessary that it is self-operating and self-correcting and requiring little or no human intervention. This approach leads to the creation of a completely new transport model, where transport planning procedures move away from trend analysis to a behavioural modelling of travel. Thus, there is a need to create and manage extensive databases on mobility, transport, user attitudes, and behavioural change. This is crucial for a better understanding of transport issues and individual customer needs.

Cargo Flow in the City

The needs of today's customers mean that there are increasing demands for flexibility of supply. The greatest increase in urban transport is observed in car transport, which provides this flexibility to the greatest extent. This concerns the transport of goods for wholesale and retail trade, service companies, public organisations, supply and purchase, transport of consumers, waste collection and disposal, and transport of supply materials to production companies located in cities and export of finished products produced in these plants. (Szołtysek, 2009a, b, c). In reality, it is difficult to precisely define urban freight flow. However, urban freight transport certainly plays an important role in sustainable urban development. He and Haasis (2020) pointed out that an urban freight transport strategy should be embedded in a comprehensive sustainable development strategy with a long-term perspective (about 20 to 30 years). Although urban freight transport plays an essential role in meeting the needs of citizens, it contributes significantly to unsustainable environmental, economic, and social impacts.

Urban freight transport is a dynamic system of high complexity, due to, inter alia, a large number of actors with different needs (logistics service providers, shippers, receivers, city and regional authorities, inhabitants and visitors, etc.), numerous constraints (e.g. traffic regulations), fragmentation of freight flow, which limits the efficiency of transport operations, new sensitive links in the supply chain (transhipments, customer contacts, last-mile deliveries, etc.), and the risk of conflicts between the expectations of different stakeholders (e.g. residents and carriers) (Iwan, 2013a, b). This system is mainly characterised by (Szołtysek, 2009a, b, c; Iwan, 2013a, b):

- heterogeneity, manifested by the multiplicity of elements and subsystems involved in its functioning, with a particular focus on diverse stakeholder groups with often different expectations and objectives,
- variety of functions and processes within each subsystem,
- the existence of numerous technical, economic, organisational, legal, social, cultural, and environmental links,
- hierarchy of subsystems,
- uniqueness, resulting from having a potential with individual parameters and operating in a specific and particular configuration of the environment,
- variability linked to the possibility of changing the number, types, and size of elements and the links between them,
- nonlinearity, leading to possible economies of scale in individual subsystems,
- spatial, area, segmental, and point irregularities,
- differentiated control, carried out at different levels, and with different degrees of impact,
- adaptability, manifested by the capacity for self-organisation and learning.

One of the key factors necessary to manage the increasing disorder within urban freight transport, and to make it sustainable, is the efficient management of goods flows between the different parties involved in its operation to a greater or lesser extent. The lack of data on freight flows, their direction, structure, etc., contributes to the difficulty to manage and direct them in a way that limits their negative impact on the urban organism, in particular on the environment and on inhabitants. These aspects have been taken into account in many European projects aimed at developing sustainable urban freight transport, such as BESTUFS/BESTUFS II, CITYLOG, CITYMOVE, SUGAR, or C-LIEGE. Moreover, documents indicating directions for the development of transport systems in Europe pay attention to these issues. They are highlighted, among others, in the "Action Plan on Urban Mobility" published by the European Commission (Iwan, 2015).

In all considerations of urban freight transport, stakeholders come to the fore. In the literature, the approach to classifying urban logistics stakeholders is not homogeneous. According to Ogden's (1992) classification of urban freight transport stakeholders (business view of urban logistics), only three actors were distinguished: institutions/persons sending and receiving freight carriers and forwarders. This classification included only private institutions/persons leaving out local government. On the other hand, local government and more precisely local authorities appeared in the classification developed by Munzuri et al. (2005), which included the following stakeholder groups: transport companies/logistics operators (all companies that deliver cargo to the urban area, including self-delivery), institutions/persons receiving cargo and local authorities (mainly responsible for the implementation of urban freight transport regulations).

As Kiba-Janiak (2018a, b) points out, in both classifications presented, stakeholders are shown rather narrowly focusing on institutions/persons that directly influence urban freight transport. A slightly broader view of urban logistics stakeholders is presented by Taniguchi et al. (2001), who added residents to the list developed by Muñuzuri et al. (2005), but only those who influence urban freight transport (e.g. by making online purchases). Anand et al. (2012), on the other hand, distinguished two groups of urban freight transport stakeholders:

- representing the public sector—otherwise known as administrators covering all managers in the city (inter alia: city authorities, directors of departments overseeing traffic and infrastructure in the city, railway, port managers in the city, etc.),
- representing the private sector (inter alia: manufacturers, shippers, suppliers, hauliers, truck drivers, cargo receivers, shopkeepers, etc.).

This is also the approach taken in this monograph.

The complexity of the urban freight transport system is due not only to the number of different stakeholders, but also to the variety of subsystems that make up the urban freight transport system. Kiba-Janiak (2017) distinguishes subsystems: delivering cargo to the city, sending cargo out of the city, transporting and storing waste, and storing cargo. The proposal of such separation of subsystems of urban freight transport is due to the fact that the issues concerning waste management are completely different from return management, therefore they should be separated in separate subsystems. In this monograph, the authors are in favour of this concept and waste management has been separated as a separate subsystem of city logistics. Returns, on the other hand, can be classified both as a subsystem of supplying goods to the city and sending goods out of the city. The storage of freight, which is important from the perspective of urban planning, can also be included in these areas. In some cities, local authorities make storage space available to companies to consolidate freight in the so-called last mile. Therefore, singling out this area in the freight flow subsystem may motivate the local authority to support urban logistics stakeholders in this respect. The freight flow subsystem can also take into account the transit traffic taking place within the city. However, this is the subsystem in which it is most difficult to control and coordinate transport, and the municipality can at most introduce some restrictions in this area by means of regulations (Taniguchi et al., 1999).

All these subsystems are undergoing dynamic changes. Changes in urban freight demand are mainly due to factors such as growing population, concentration of industrial production, increased demand due to e-commerce, longer daily activity periods, larger city areas, and increasing customer demand. Due to the impact of these factors, new problems arise, including difficulties in transport coordination and conflicting expectations of all stakeholders in the urban transport system (Mężyk & Zamkowska, 2016). A key factor that shapes urban freight flows in recent times is the pandemic. Consumer purchasing behaviour is changing, and some of these changes will be perpetuated irrespective of the period of impact of COVID 19. These changes are particularly related to the development of e-commerce together with innovations in sending and receiving parcels.

The planning of urban freight transport systems is invariably a considerable challenge for those responsible. In spite of a good theoretical basis, even in developed countries with mature local governance systems, freight transport escapes the tried and tested patterns. There are few countries where systematic efforts have been made to integrate freight planning with spatial, economic, and social planning.

Cargo transport problems are particularly evident in large cities and agglomerations. They are a specific study area. On the one hand, in their role as economic and administrative centres, they concentrate the demand for goods in a limited area, where all disadvantages associated with transport activities are concentrated. On the other hand, the concentration of freight, receivers, carriers, and infrastructure provides an opportunity to effectively influence these elements of the transport system by means of rationally selected measures.

In the urban freight management and planning system, it is therefore necessary to take into account the complex structure of the actors, their interests, and decision-making factors. Local authorities have a special role to play. Their most important asset with respect to other actors is their neutrality (Lindholm, 2012). This should be a motivating factor for them to engage in planning and implementing new solutions and to act as a coordinator of the system. The structure of the relationship between actors and regulators in the implementation of freight transport measures is shown in Fig. 2.5.

Traditionally, local authorities act as the author of transport and spatial policy, infrastructure provider, and traffic managers by various means

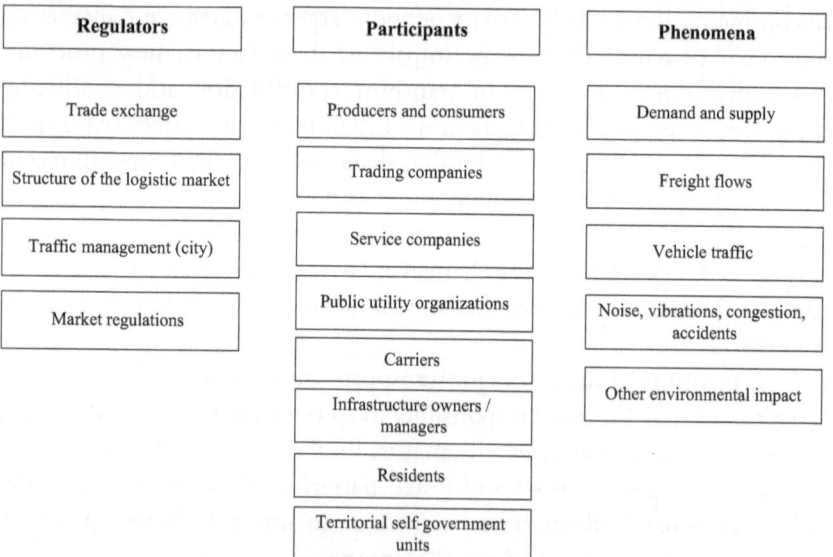

Fig. 2.5 Regulators and participants in urban freight flows. Source: compiled on the basis of: Visser et al. (1999)

(regulation of infrastructure demand) (Kaszubowski, 2013). The juxta-position should be extended to other actors, for example, carriers and logistics operators as well as shippers and receivers. This should make it possible to simultaneously influence the two basic processes in urban freight transport, that is, both the volume and structure of vehicle traffic and the volume of freight flow. Such an approach is indispensable in view of the complexity of the problems involved in urban freight transport. In functional terms, these most often include (Allen et al., 2007):

- problems related to truck and delivery vehicle traffic, concerning congestion, excessive demand on infrastructure, and negative impact on its quality,
- problems associated with existing regulations, which include all forms of weight restrictions on vehicles, delivery time restrictions, exclusion zones, etc.,
- difficulties related to parking and loading operations of heavy goods vehicles, where the lack of a sufficient number of well-planned dedicated loading places, problems with the misuse of existing places by unauthorised users, or existing regulations related to parking time are particularly noticeable,
- problems arising between the customer (receiver) and the carrier (supplier), such as waiting for the preparation and loading of the shipment, lack of appropriate documents, difficulties in finding the receiver, the need to take into account the specific requirements of the customer, such as delivery time,
- unreliability of carriers is due to the strong pressure to reduce operating costs in a very competitive market, which often leads to reducing investment in rolling stock or offering services with parameters that do not match the actual capacity of the carrier; this results in poor quality of service and difficulties in establishing long-term customer relationships.

The problems presented should be seen in the light of the objectives that a sustainable urban freight transport system must fulfil. An efficient urban freight transport system should have the following characteristics, taking into account the main user groups and regulatory processes shown in Fig. 2.5 (Behrends et al., 2008):

- guarantee accessibility to all modes of transport forming part of the system, without impeding socio-economic processes,
- contribute to increasing the cost and energy efficiency of transport operations, taking into account all external costs,
- contribute to the value and attractiveness of urban space by reducing land taking for transport functions, improving transport safety, and removing factors that limit the mobility of residents,
- reduce the negative impact of transport activities on the environment, both natural and man-made.

Urban freight optimisation projects carried out in many countries around the world allow us to present a classification of solutions that are used to meet the requirements mentioned above. It should be noted that these solutions are used on different scales and with very different effects. Table 2.3 presents a classification of solutions that can be used by local authorities to address the problem under consideration.

In most cases, cities decide to implement the chosen tools based on their knowledge of local problems and by using the available information, also in the form of experience transfer from other cities. These areas are developed in studies on urban freight flows as different possibilities to coordinate these flows. The problems of routing delivery vehicles with time windows were studied by Li et al. (2021). The authors analysed the effect of time window length and compared simultaneous and separate delivery modes. The capacity problems of vehicle routes with alternative: delivery—collection—time windows were measured by Sitek et al. (2021). Johansson and Björklund (2017), on the other hand, addressed the demand of retail shops for urban consolidation centres (UCCs). UCCs are often established to improve retail shop services and potentially reduce costs. Zunder and Marinov (2011) confirmed that in the construction sector, urban consolidation centres can be successfully used to improve logistics operations.

Table 2.3 Potential solutions to improve urban logistics in the freight flow subsystem

Category	Group	Particular solutions
Infrastructure and town planning	Transshipment hubs	City terminals
		Urban transfer centres
		Rail or sea terminals
		Freight clusters (freight villages, or logistics terminals)
		Bypasses
		Acceleration/deceleration lanes (truck climbing lanes)
		Removal of geometric constraints at intersections
	Parking places	Loading area (freight, parking, and loading zones)
		On-street parking and loading
		Planned location of parking spaces
		Conditional use of places not previously used by delivery vehicles, e.g., bus lanes or taxi ranks
		Vehicle parking reservation systems
		Timeshare of parking space
		Upgrade parking areas and loading docks
		Truck stops/parking outside of metropolitan areas
	Building regulations	Indoor loading bay
		Conditional use of private parking spaces
		Mini storage in the form of maintenance-free lockers
		Enhanced building codes
		Staging areas
Access restrictions	Restrictions in the defined area	Vehicle size and weight restrictions
		Engine-related restrictions
		Access rules to pedestrian areas
		Entry fees
		Exclusion of the area from private vehicle traffic
		Loading and parking restrictions
		Truck routes
		Low emission zone
		Load factor restriction
		Exclusive truck lanes (dedicated truck lanes)
	Time restrictions	Designated delivery time (time windows)
		Overnight deliveries
		Adequate rotation at delivery points
		Peak-hour clearways

(continued)

Table 2.3 (continued)

Category	Group	Particular solutions
Traffic management	Regulation scope	Regulation of delivery times and area Harmonisation of legislation Emission standards Low noise delivery programmes Traffic control
	Information	Virtual reservation of parking spaces for deliveries

Source: compiled from: Munzuri et al. (2005)

Logistical Aspects of Waste Management

Waste management logistics is a key factor in environmental protection and the rational use of materials and raw material resources and currently represents the fourth, equally important element of the logistics system along with supply, distribution, and production logistics.

The concept of waste management logistics has evolved over the last decades, its evolution being highlighted by various authors. The first formally accepted definition of logistical management of waste streams, endorsed in the early 1990s by the Council of Supply Chain Management Professionals, is that the role of logistics in recycling, waste management, and hazardous materials management refers to all issues related to logistical activities undertaken to reduce raw material consumption, recycling, substitution, material reuse, and management (Grabara & Man, 2014). It is important to highlight here the strong emphasis placed on the elements of waste recovery and reuse as activities embedded in the logistical management of waste streams. It is now recognised that the logistical management of waste streams is the process of moving products to their final destination to get the right value or action for them. The term waste can also be applied to products in logistic waste streams, which in most cases are treated as defective or waste products (Grabara & Man, 2014). Hence, many authors interchangeably define waste stream logistics and reverse logistics (Reverse Logistics, RL) (Carrasco, 2010; Fernández, 2003; De Brito & Dekker, 2004; Rogers & Tibben-Lembke, 1999). However, defining reverse logistics as part of reverse supply chain, responsible for collecting products from the market and selling them or processing them to recover part of the market value or disposing of them (Rogers & Tibben-Lembke, 1999), it is an important part of the return management process. It should be emphasised that the subjects of reverse logistics are the waste flows (subject to their broad understanding) and information related to those flows (Szołtysek, 2009a, b, c), apart from that, material flows related to the withdrawal of valuable products from the logistics system and related to the process of repair, complaints are also included. Therefore, the authors believe that differences in defining the concept of reverse logistics and waste logistics (streams) should be clearly indicated.

An integrated waste management logistics system should be constructed taking into account functional areas (Bril et al., 2016) covering activities that are related to waste generation, transport to waste management facilities, storage, economic use, and disposal (areas in the real sphere) and due to activities that include regulation and control (areas in the regulatory sphere). The basic determinants of the functioning of an integrated waste management logistics system are considered (Bril & Rydygier, 2016; Jąderko & Białecka, 2013):

- quantity, nature, and spatial distribution of waste,
- degree of regularity and dynamics of waste generation,
- principles of environmental protection,
- limited capacity and throughput of installations, equipment, and facilities that are part of the logistics chain,
- the spatial capacity of the areas covered by unitary waste management and the availability of facilities using Best Available Techniques (BAT) for waste disposal,
- spatial and urban factors: the structure and shape of the settlement network of the region, the possibility of locating system facilities, communication routes, the spatial structure of economic activity, etc.,
- general standards and local (local and regional) requirements for acceptable environmental pollution,
- updates to the laws, regulations, and provisions that form the basis of the system,
- global trends in waste management concepts.

Table 2.4 Areas of logistical support for waste management

Areas of logistic support	Description
Ecological	Support solving problems related to the negative impact of human activity on the environment. Ensures the efficient flow of waste while respecting the principles of protection of the human environment (conservation of resources and reduction of pollution)
Economic	Creates logistics chains linking waste generation points to treatment points to ensure a cost-effective flow of waste with associated information (cost reduction)
Organisational	It is the process of planning, implementing, and controlling the efficient flow of waste for recovery or proper management. It integrates waste flow in time and space, while optimising costs and minimising environmental impacts. It leads to an improved level of service for the collection, transport, and treatment of waste

Source: Kalisiak-Mędelska (2017)

Planning of rational waste management meeting restrictive ecological requirements requires the use of various and appropriate solutions in this respect (technical, organisational, legal, logistic, financial, etc.). Logistics, with a range of solutions to improve the management of waste flows in the city, including the elimination of irrational activities and costs, is able to meet the growing requirements of waste management and adapt to its changes (Table 2.4).

The main areas of support for waste management and logistics are economic and ecological. The former implies reducing logistics costs and striving for continuous improvement of customer service. Production waste should be treated according to the method of greening economy, according to which the secondary raw material should be treated as a specific commodity that can be put back on the market (if it has a certain value or the costs of its disposal are too high). The economic benefits should be understood as the savings that arise from the process of waste disposal or when income is obtained from the sale of recovered raw materials. The second (ecological) area manifests itself in the reduction of pollution that is generated by the disposal processes carried out. In the long term, these areas become convergent goals. Furthermore, an important area is the organisational area consisting of creating an integrated waste management system taking into account the principles of sustainable development.

Within the concept of sustainability, the contemporary concept of logistics revolves around the issue of ecological logistics (ecologistics) and is considered an effective approach to managing materials and accompanying information flows to reduce ecological and economic damage to the environment (Van Buren et al., 2016). Ecologistics contributes to preventing and eliminating the effects of negative ecodestructive impacts on the environment by transforming logistics systems, corresponding to the modern linear model of the economy, into ecologistic systems (Rudenko & Kovtun, 2020). Similarly, logistically integrated waste management requires the use of instruments of modern logistics, which makes it possible to integrate the waste management subsystem with entities managing returns or using secondary raw materials, as well as entities engaged in waste disposal. The effect of integration understood in this way may be the achievement of the required level of effectiveness of the waste management system, which implies the achievement of the desired efficiency and a satisfactory level of customer service (entity—links in the waste chain). Such defined (expected) effects of integration are possible to achieve due of the similarity to solutions applied within supply chains (Smyk, 2016). Integration of undertakings implemented within the waste chain is a necessary condition for designing a system solution and separating subsystems from the logistically integrated waste management system such as:

- waste collection subsystem,
- waste transport and storage subsystem,
- waste recovery subsystem,
- waste disposal subsystem
- information and decision-making subsystem.

The first element of an effective waste management system is properly organised waste collection by waste generators, ensuring its further smooth collection and transport. The choice of the collection system[1] should be preceded by thorough research and analysis of the inhabited and developed area. The next stage, that is, collection of waste collected by waste generators, depends on the method of collection and the type of waste collection containers used and is determined by the need to use specific collection equipment. In the waste recovery subsystem, activities focus on sorting waste collected in a given area and its proper recovery or recycling. At this point it should be emphasised that in Poland the evolution of recycling is blocked by many aspects, the most important of which are (Lutek et al., 2019):

- significant levels of waste contamination,
- substantial charges for the collection of selectively collected waste,
- high cost of recycling activities,
- changing the legal regulations in the field of environmental protection,
- fluctuations in raw material prices,
- low environmental awareness of the population.

The waste disposal subsystem involves waste whose generation could not be prevented or which could not be recovered. In this case, the waste undergoes biological, physical, or chemical transformation processes[2] so that it does not pose a threat to humans and the environment.

The waste management subsystems presented will not function properly without an appropriate information and decision-making subsystem and without the involvement of participants and stakeholders in the system. Social acceptance of waste handling methods is also an important issue. Moreover, an important aspect is also the continuous improvement of consumers' environmental awareness. Therefore, as part of the idea of ecologistics, consumers should be promoted and taught such attitudes that are conducive to environmental protection. Consumers being guided

---

[1] Possible system solutions for the collection of individual waste fractions are direct collection systems, donation systems, mixed systems, and complementary systems.

[2] Disposal of waste is carried out, inter alia, by thermal methods (incineration) and by depositing in landfills.

by such principles will be conducive to minimising waste generation, achieving appropriate levels of recovery and recycling, and reducing the amount of waste deposited in landfills (Baran & Karlewska, 2016).

References

Abduljabbar, R., Liyanage, S., & Dia, H. (2021). The role of micro-mobility in shaping sustainable cities: A systematic literature review. *Transportation Research Part D, 92*, 1–19.

Allen, J., Muñoz, J. C., & de Dios Ortúzar, J. (2019). Understanding public transport satisfaction: Using Maslow's hierarchy of (transit) needs. *Transport Policy, 81*, 75–94.

Allen, J., Thorne, G., & Browne, M. (2007). BESTUFS: Good practices guide on urban freight transport.

Anand, N., Yang, M., van Duin, R., & Tavassy, L. (2012). GenClon: An ontology for city logistics. *Expert Systems with Applications, 13*(15), 11944–11960.

Baran, J., & Karlewska, M. (2016). Teoretyczne aspekty logistycznego systemu gospodarki odpadami. *Ekonomika i Organizacja Logistyki, 1*(3), 5–17.

Barceló, J., Grzybowska, H., & Pardo, S. (2007). Vehicle routing and scheduling models, simulation and city logistics. In V. Zeimpekis, C. D. Tarantilis, G. M. Giaglis, & I. Minis (Eds.), *Dynamic fleet management concepts—systems, algorithms & case studies* (pp. 163–195). Springer.

Behrends, S., Lindholm, M., & Woxenius, S. (2008). The impact of urban freight: A definition of sustainability from an actor's perspective. *Transportation Planning and Technology, 31*(6), 2008.

Bektaş, T., Crainic, T. G., & Van Woensel, T. (2015). From managing urban freight to Smart City logistics networks. Québec.

Bektas, T., Crainic, T. G., & Van Woensel, T. (2017). From managing urban freight to smart city logistics networks. In K. Gakis & P. Pardalos (Eds.), *Networks design and optimization for smart cities* (Series on computers and operations research) (Vol. 8, pp. 143–188).

Benevolo, C., Dameri, R. P., & D'Auria, B. (2016). Smart mobility in smart city: Action taxonomy, ICT intensity and public benefits. In T. Torre, A. M. Braccini, & R. Spinelli (Eds.), *Empowering organizations* (pp. 13–28). Springer.

Borys T., Rogala P. (red.) (2009). Jakość życia na poziomie lokalnym—ujęcie wskaźnikowe, Program Narodów Zjednoczonych ds Rozwoju .

Bosona, T. (2020). Urban freight last mile logistics—challenges and opportunities to improve sustainability: A literature review. *Sustainability, 12*, 8769.

Bril, J., Łukasik, Z., & Rydygier, E. (2016). Aspekty logistyczne gminnych systemów gospodarowania odpadami komunalnymi. *Autobusy, 6*, 1251.

Bril, J., & Rydygier, E. (2016). Effective municipal waste management as a challenge for self-government municipalities. *International Journal of Engineering and Advanced Research Technology, 2*(1).

Bulkeley, H., & Castán Broto, V. (2013). Government by experiment? Global cities and the governing of climate change. *Transactions of the Institute of British Geographers, 38*(3), 361–375.

Carrasco, R. A. (2010). Management model for closed-loop supply chains of reusable articles. PhD thesis, Universidad Politécnica de Madrid, Madrid, Spain, September 2010.

Cavoli, C. (2021). Accelerating sustainable mobility and land-use transitions in rapidly growing cities: Identifying common patterns and enabling factors. *Journal of Transport Geography, 94*, 1–13.

Chamier-Gliszczyński, N. (2011). Zrównoważony miejski system transportowy. *Autobusy, 5*, 88–93.

Cichosz, M. (2014). Transport pasażerski w miastach. In Innowacje w zarządzaniu miastami w Polsce. editors. M. Bryx, Oficyna Wydawnicza SGH, Warszawa.

Crainic, T. G., Perboli, G., & Ricciardi, N. (2020). City logistics. In T. G. Crainic, M. Gendreau, & B. Gendron (Eds.), *Network design with applications in transportation and logistics*. Springer. Forthcoming.

Crainic, T. G., Ricciardi, N., & Storchi, G. (2009). Models for evaluating and planning city logistics systems. *Transportation Science, 43*(4), 432–454.

De Brito, M. P., & Dekker, R. (2004). A framework for reverse logistics. In R. Dekker, M. Fleischmann, K. Inderfurth, & L. N. Van Wassenhove (Eds.), *Reverse logistics. Quantitative models for closed-loop supply chains* (pp. 3–28). Springer.

Debnath, A., Chin, H. C., Haque, M. M., & Yuen, B. (2014). A methodological framework for benchmarking smart transport cities. *Cities, 37*, 47–56.

Dillman, K., Czepkiewicz, M., Heinonen, J., Fazeli, R., Árnadóttir, Á., Davíðsdóttir, B., & Shafiei, E. (2021). Decarbonization scenarios for Reykjavik's passenger transport: The combined effects of behavioural changes and technological developments. *Sustainable Cities and Society, 65*, 102614.

Fernández, I. (2003). The concept of reverse logistics. A review of the literature. *Reverse Logistics Digit Magazine, 2003*(58), 47–49.

Fontaine, P., Crainic, T. G., Jabali, O., & Rei, W. (2021). Scheduled service network design with resource management for two-tier multimodal city logistics. *European Journal of Operational Research, 294*(2), 558–570.

Gianessi, P., Alfandari, L., Létocart, L., & Calvo, R. W. (2016). The multicommodity-ring location routing problem. *Transportation Science, 50*(2), 541–558.

Gillingham, R., & Reece, V. J. S. (1980). Analytical problems in the measurement of the quality of life. *Social Indicators Research, 1*, 2.

Glazer, W. (2012). Cross-national comparisons of quality of life in developed nations, including the impact of globalization. In K. C. Land, A. C. Michalos, & J. M. Sirgy (Eds.), *Handbook of social indicators and quality of life research*. Springer.

Grabara, J., & Man, M. (2014). Assessment of logistic outlays in industrial solid waste management. *Social Sciences and Education Research Review, 1*(2), 11–21.

He, Z., & Haasis, H. D. (2020). A theoretical research framework of future sustainable urban freight transport for smart cities. *Sustainability, 12*, 1975.

Hebel, K., & Wyszomirski, O. (2016). Ewolucja postulatów przewozowych dotyczących podróży miejskich mieszkańców Gdyni w świetle wyników badan marketingowych z lat 1985-2015. *Problemy Transportu i Logistyki, 3*, 63–71.

Holguin-Veras, J., Amaya, J., Sasnhez-Diaz, I., & Browne, M. (2021). State of the art and practice of urban freight management, part 1: Infrastructure, vehicle—related and traffic operations. *Transportation Research Part A Policy and Practice, 137*, 360–382.

Iwan S., (2013a). *Wdrażanie dobrych praktyk w obszarze transportu dostawczego w miastach* (p. 56). Wydawnictwo Naukowe Akademii Morskiej, Szczecin.

Iwan, S. (2013b). *Wdrażanie dobrych praktyk w obszarze transportu dostawczego w miastach* (p. 20). Wydawnictwo Naukowe Akademii Morskiej.

Iwan, S. (2015). Analiza wybranych miast polskich pod kątem funkcjonowania miejskiego transportu towarowego, Logistyka.

Jąderko, K., & Białecka, B. (2013). Wybrane problemy budowy systemu gospodarki odpadami komunalnymi w świetle nowych przepisów. *Systemy Wspomagania w Inżynierii Produkcji, z., 3*(5), 135–144.

Jain, D., & Tiwari, G. (2016). How the present would have looked like? Impact of non-motorized transport and public transport infrastructure on travel behavior, energy consumption and CO_2 emissions—Delhi, Pune and Patna. *Sustainable Cities and Society, 22*, 1–10.

Janczewski, J. (2020). Mikromobilność w systemie transportowym miasta. *Przedsiębiorczość – Edukacja, 16*(1), 257–274.

Johansson, H., & Björklund, M. (2017). Urban consolidation centres: Retail stores' demands for UCC services. *International Journal of Physical Distribution and Logistics Management, 47*, 646–662.

Kalisiak-Mędelska, M. (2017). Logistyka a gospodarka odpadami w mieście. *Studia Miejskie, 27*, 64.

Kaszubowski, D. (2013). Planowanie systemu transportu ładunków w miastach. *Autobusy, 3*.

Kiba-Janiak, M. (2017). Opportunities and threats for city logistics development from a local authority perspective. *Journal of Economics and Management, 28*(2), 23–39. ISSN 1732-1948Wydawnictwo Uniwersytetu Ekonomicznego w Katowicach.

Kiba-Janiak, M. (2018a). *Logistyka w strategiach rozwoju miast*. Wydawnictwo Unwersytetu Ekonomicznego we Wrocławiu.

Kiba-Janiak, M. (2018b). *Logistyka w strategiach rozwoju miast* (pp. 42–44). Wydawnictwo UE we Wrocławiu.

Kiba-Janiak, M., & Cheba, K. (2011). An assessment of individual transport in the aspect of quality of life on the example of selected medium sized cities. *Total Logistics Management, 4*, 77–88.

Kiba-Janiak, M., & Cheba, K. (2014). How local authorities are engaged in implementation of projects related to passenger and freight transport in order to reduce environmental degradation in the city. *Procedia - Social and Behavioral Sciences, 151*, 127–141.

Kiba-Janiak, M., & Witkowski, J. (2014). *Modelowanie logistyki miejskiej* (p. 11). PWN.

Kolman, R. (2009). *Kwalitologia. Wiedza o różnych dziedzinach jakości*. Placet.

Krygsman, S., Dijst, M., & Arentze, A. (2004). Multimodal public transport: An analysis of travel time elements and the interconnectivity ratio. *Transport Policy, 11*, 265–275.

Krysiuk, C., & Nowacki, G. (2016). Miasto, element systemu transportowego kraju. *Autobusy, 10*, 124–130.

Kwan, S. C., & Hashim, J. H. (2016). A review on co-benefits of mass public transportation in climate change mitigation. *Sustainable Cities and Society, 22*, 11–18.

Li, W., Li, K., Kumar, P. N. R., & Tian, Q. (2021). Simultaneous product and service delivery vehicle routing problem with time windows and order release dates. *Applied Mathematical Modelling, 89*, 669–687.

Lindholm, M. (2012). How local authority decision makers address freight transport in urban areas. *Procedia Social and Behavioral Sciences, 39*.

Liu, B. C. (1975a). Differential net migration rates and the quality of life. *The Review of Economics and Statistics, 57, 122*.

Liu, B. C. (1975b). Metropolia quality of life indicators in the US. *Warszawa*.

Lutek, W., Pastuszak, Z., & Banaś, J. (2019). *Smart City Innowacyjny system zarządzania logistyką zwrotną w gospodarce odpadami komunalnymi* (pp. 125–128). UMCS.

Matusiewicz, M. (2019). Towards sustainable urban logistics: Creating sustainable urban freight transport on the example of a limited accessibility zone in Gdansk. *Sustainability, 11*, 3879.

Mężyk, A., & Zamkowska, S. (2016). Problemy obsługi logistycznej miast w zakresie dostaw ładunków. *Autobusy, 6*, 1445–1448.

Mogaji, E., & Nguyen, N. P. (2021). Transportation satisfaction of disabled passengers: Evidence from a developing country. *Transportation Research Part D, 98*, 1–15.

Mouwen, A. (2015). Drivers of customer satisfaction with public transport services. *Transportation Research Part A, 78*, 1–20.

Muñuzuri, J., Larrañeta, J., Onieva, L., & Cortés, P. (2005). Solutions applicable by local administrations for urban logistics improvement. *Cities, 22*(1), 15–28.

Munzuri, J., Larraneta, J., Onieva, L., & Cortes, P. (2005). Solutions applicable by local administrations for urban logistics improvement. *Cities, 22*(1), 15–28.

Oeschger, G., Carroll, P., & Caulfield, B. (2020). Micromobility and public transport integration: The current state of knowledge. *Transportation Research Part D, 89*, 1–21.

Ogden, K. W. (1992). *Urban goods movement: A guide to policy and planning.* Ashgate Publishing Company.

Paiva, S., Ahad, M., Tripathi, G., Feroz, N., & Casalino, G. (2001). Enabling technologies for urban smart mobility: Recent trends, opportunities and challenges. *Sensors, 21*(2143), 3–41.

Regmi, M. B. (2020). Measuring sustainability of urban mobility: A pilot study of Asian cities. *Case Studies on Transport Policy, 8*, 1224–1232.

Rocha, N., Santinha, G., Dias, A., Rodrigues, C., Rodrigues, M., & Queirós, A. (2021). A systematic literature review of smart cities' information services to support the mobility of impaired people. *Procedia Computer Science, 181*, 182–188.

Rogers, D. S., & Tibben-Lembke, R. S. (1999). *Going backwards reverse logistics trends and practices* (1st ed., pp. 1–280). Council of Reverse Logistics.

Ros-McDonnell, L., de-la-Fuente-Aragón, M. V., Ros-McDonnell, D., & Cardós, M. (2018). Analysis of freight distribution flows in an urban functional area. *Cities, 79*, 159–168.

Rudenko, S. V., & Kovtun, T. A. (2020). Greening of logistics as a direction of realization of the concept of sustainable development. *Project and Logistics Management: New Knowledge Based on Two Methodologies, 3*, 7–24.

Rudke, A., Martins, J., dos Santos, A., Silva, P., da Silva Caldana, N., Souza, V., Alves, R., & de Almeida Albuquerque, T. (2021). Spatial and socio-economic analysis of public transport systems in large cities: A case study for Belo Horizonte, Brazil. *Journal of Transport Geography, 91*, 1–11.

Rzeczyński, B. (2007). *Logistyka miejska. Propedeutyka pierwszy polski wykład* (p. 23). Instytut Inżynierii Zarządzania Politechniki Poznańskiej.

Savelsbergh, M., & Van Woensel, T. (2016). City logistics: Challenges and opportunities. *Transportation Science, 50*(2), 579–590.

Semanjski, I., & Gautama, S. (2019). A collaborative stakeholder decision-making approach for sustainable urban logistics. *Sustainability, 11*, 234.

Sethi, M., Lamb, W., Minx, J., & Creutzig, F. (2020). Climate change mitigation in cities: A systematic scoping of case studies. *Environmental Research Letters, 15*(9), 1–16.

Shafique, M., Azam, A., Rafiq, M., & Luo, X. (2020). Evaluating the relationship between freight transport, economic prosperity, urbanization, and CO2 emissions: Evidence from Hong Kong, Singapore, and South Korea. *Sustainability, 12*, 10664.

Sitek, P., Wikarek, J., Rutczynska-Wdowiak, K., Bocewicz, G., & Banaszak, Z. (2021). Optimization of capacitated vehicle routing problem with alternative delivery, pick-up and time windows: A modified hybrid approach. *Neurocomputing, 423*, 670–678.

Skultety, F., Benova, D., & Gnap, J. (2021). City logistis as an imperative smart city mechanism: Scrutiny of lustered EU27 capitals. *Sustainability, 13*, 3641.

Smyk, S. (2016). Logistycznie zintegrowana gospodarka odpadami jako współczesne wyzwanie społeczno-gospodarcze. *Logistyka, 1*, 400.

Strychul, M., Khomeriki, O., Mykhailych, O., Yahodzinskyi, S., Romanenko, Y., & Perelyhin, T. (2021). Functioning of public transport in the social space of the city. *Transportation Research Procedia, 54*, 610–616.

Sun, G., & Lau, C. (2021). Go-along with older people to public transport in high-density cities: Understanding the concerns and walking barriers through their lens. *Journal of Transport & Health, 21*, 1–16.

Szołtysek, J. (2005). *Logistyczne aspekty zarządzania przepływami osób i ładunków w miastach.* Uniwersytet Ekonomiczny w Katowicach.

Szołtysek, J. (2009a). *Logistyczne aspekty zarządzania przepływami osób i ładunków w miastach.* Wydawnictwo AE Katowice.

Szołtysek J. (2009b). Logistyka zwrotna. In: D. Kisperska-Moroń, S. Krzyżaniak (red.), Logistyka, Biblioteka Logistyka, Poznań, s. 436.

Szołtysek, J. (2009c). *Logistyczne aspekty zarządzania przepływami osób i ładunków w miastach, Prace naukowe Akademii Ekonomicznej im.* Karola Adamieckiego.

Szołtysek, J. (2010). Logistyka miasta. Geneza, istota, perspektywy. *Logistyka, 5,* 8.

Szołtysek, J. (2016). *Logistyka miasta* (p. 49). PWE.

Szołtysek, J., & Otręba, R. (2015). Mobilność a jakość życia mieszkańców miast Górnego Śląska. *Transport Miejski i Regionalny, 11,* 21–26.

Szymczak, M. (2006). O istocie i funkcjach logistyki miejskiej, w: Współczesne kierunki rozwoju logistyki, E. Gołembska [red. nauk.], PWE, Warszawa., s. 84–85.

Taniguchi, E. (2015). City logistics for sustainable and liveable cities. *Green Logistics Transportation, 151,* 49–60. https://doi.org/10.1007/978-3-319-17181-4_4

Taniguchi, E., Thompson, R. G., & Yamada, T. (1999). Modelling city logistics (w): City logistics I, (red.) E. Taniguchi, R. G. Thompson, Institute of System Science Research, Kyoto, s. 3–37.

Taniguchi, E., Thompson, R. G., & Yamada, T. 2003, Visions for city logistics in logistics systems for sustainable cities. *Proceedings of the 3th international conference on city logistics, Madeira Portugal, 25–27 June 2003,* Amsterdam: Elsevier.

Taniguchi, E., & van der Heijden, R. E. C. M. (2000). An evaluation methodology for city logistics. *Transport Reviews, 20*(1), 65–90.

Taniguchi, E., Thompson, R. G., Yamada, T., & van Duin, J. H. R. (2001). *City logistics: Network modelling and intelligent transport systems.* Pergamon.

Tirachini, A., Chaniotakis, E., & Abouelela, M. (2020). The sustainability of shared mobility: Can a platform for shared rides reduce motorized traffic in cities? *Transportation Research Part C, 117,* 1–15.

Tomaszewska, E. (2015). Inteligentny system transportowy w mieście na przykładzie Białegostoku. Zeszyty Naukowe Uniwersytetu Szczecińskiego. *Problemy zarządzania, finansów i marketing, 41*(2), 317–329.

Tundys B. (2008a). Logistyka miejska.

Tundys, B. (2008b). *Logistyka miejska. Koncepcje, systemy, rozwiązania* (p. 145). Difin.

Tundys, B. (2013). *Logistyka miejska. Teoria i praktyka.* Difin.

Van Buren, N., Demmers, M., van der Heijden, R., & Witlox, F. (2016). Towards a circular economy: The role of Dutch logistics industries and governments. *Sustainability, 8*(7), 647.

Visser, J., Binsbergen, A., & Nemoto, T. (1999). *Urban freight transport policy and planning.* First International Symposium on City Logistics, Cairns, Australia.

Wiktorowska-Jasik, A. (2014). *Analiza wybranych elementów systemu logistycznego Szczecina* (p. 20). Transport Miejski i Regionalny.

Wimbadi, R., Djalante, R., & Mori, A. (2021). Urban experiments with public transport for low carbon mobility transitions in cities: A systematic literature review (1990–2020). *Sustainable Cities and Society, 72*, 1–17.

Wojewódzka-Król, K., & Załoga, E. (red.). (2016). Transport. Nowe wyzwania. PWN

Wyszomirski, O. (2008). *Transport miejski. Ekonomika i organizacja.* Wydawnictwo Uniwersytetu Gdańskiego.

Yang, L. (2018). Modeling the mobility choices of older people in a transit-oriented city: Policy insights. *Habitat International, 76*, 10–18.

Zawieska, J., & Pieriegud, J. (2018). Smart city as a tool for sustainable mobility and transport decarbonisation. *Transport Policy, 63*, 39–50.

Zeliaś, A. (2004). Taksonomiczna analiza przestrzennego zróżnicowania poziomu życia w Polsce w ujęciu dynamicznym. Wyd. Akademii Ekonomicznej w Krakowie, Kraków, s. 23–34.

Zimon, G., & Gosik, B. (2015). Ocena logistyki miejskiej w zakresie transportu zbiorowego na przykładzie Tomaszowa Mazowieckiego i Rzeszowa. Modern. *Management Review, 22*(2), 197–209.

Zunder, T., & Marinov, M. (2011). Urban freight concepts and practice: Would a traditional UCC scheme work? *Transportation Problem, 6*, 87–95.

3

Logistic Maturity of Cities

The Context of Maturity

Management science increasingly uses the term maturity in various contexts. It refers to phenomena and processes that can contribute to the development of an organisation, and the process of achieving maturity remains linked to the improvement of skills (Skrzypek, 2013). The concept of maturity refers to *the state of being complete, perfect, or ready* (Wibowo & Waluyo, 2015). It implies a certain level of skill (excellence) and can be considered as management, process, project, technology, quality, knowledge, culture, and maturity. However, an analysis of the world literature indicates that the dominant research area in this sphere is organisational maturity, considered in the context of a process approach (Kerzner, 2001; Harmon, 2014; Rosemann & de Bruin, 2005; Kenny, 2006; Mutafelija & Stromberg, 2008; Hüner et al., 2009; Rohloff, 2009; Pöppelbuß & Röglinger, 2011; Albliwi et al., 2014; de Boer et al., 2015; Dijkman et al., 2016; Tarhan et al., 2016; Wolniak, 2019). The concept of maturity is most often presented in the context of **maturity models**. The idea of a maturity model was first described by Nolan (1973) and Crosby (1979). It can represent a set of various tools and practices that

© The Author(s), under exclusive license to Springer Nature Switzerland AG 2022
M. Kramarz et al., *Urban Logistics in a Digital World*,
https://doi.org/10.1007/978-3-031-12891-2_3

enable the assessment of an organisation's management competencies (OGC, 2007), as well as the improvement of key factors leading to the achievement of its objectives (Looy, 2014). Maturity models allow for a holistic view of the organisation and a comprehensive assessment in terms of meeting the key expectations set by various requirements (e.g. legal acts, management concept, assumptions, or internal company arrangements) and stakeholders (customers, counterparties, employees, society, etc.). The maturity model should be constructed in such a way that it delineates the paths of improvement, ultimately leading to the improvement of the economic results obtained and the competitive position (Kosieradzka, 2016). Organisational maturity models used by business organisations can be divided according to the particular areas of the organisation's activity to which the model applies, hence process maturity models can be distinguished (e.g. Business Process Maturity Model— BPMM,[1] Process and Enterprise Maturity Model—PEMM), project maturity models (e.g. Project Management Maturity Model—PMMM, PRINCE2 Maturity Model—P2MM), quality maturity models (e.g. Quality Management Maturity Grid—QMMG, EFQM excellence model—developed by the European Foundation for Quality Management) and others.

The identification of many different approaches to the issue of maturity demonstrates the great variety and wide applicability of maturity models available in the literature. Despite the identified differences in their interpretation, one gets the impression that at the same time there are many similarities within their applications. Referring to the study of maturity models conducted by Kosieradzka and Smagowicz (2017), it can be stated that from the point of view of the structure of models, the vast majority of them are characterised by the number of maturity levels oscillating between 4 and 6 (18 models studied), with 5 maturity levels prevailing in most of them. In the structure of individual models, it is most common to distinguish areas to which general and specific objectives have been assigned, as well as practices (methods and techniques)

---

[1] This model, developed by the international organisation Object Management Group (OMG), identifies six areas of assessment, with a set of guidelines developed for each to qualify for the appropriate of five maturity levels.

enabling their achievement. Moreover, all models described above were characterised by a wide range of uses, including both the identification of the current and target state of the organisation. The creators of the model used the concept of continuous improvement and enabled organisations to answer the question "where is the organisation now", but also to determine future directions of development. In terms of the way information was collected to apply the models, in nine of the cases analysed, it was self-assessment, allowing the creation of a picture of the current state of the organisation by the employees themselves, who have detailed knowledge about the functioning of the organisation. In the remaining cases, questionnaire surveys and direct observation were used, as well as an interview in the case of assessments carried out by external experts. The research clearly shows that the way of measuring results in most models is similar and consists in determining the degree to which the criteria for particular areas are met on a percentage scale (14 models studied). Scoring and weighting are also used, especially in situations where the importance of individual areas varies. The indicated similarities within the analysed models indicate that their authors rely mainly on previous studies, hence it can be considered that they are the foundation of the developed models. It should also be emphasised that each model should be developed with reference to key factors of importance for the development of a given organisation. Because the aim of all maturity models is to lead to continuous improvement and enhancement of the organisation. The process of continuous improvement should condition the increase of the level of maturity: managerial and organisational, praxeological, sociocultural, technical, and technological or environmental (Table 3.1).

Maturity models can therefore be considered as practically used tools to assess processes and organisations, and the multitude of their applications convinces of their important role in the management sphere, in areas such as process management, knowledge management, project management, sustainability, risk management, supply chain, education, public sector, construction, service management, medical sector, human resources management, and product lifecycle management, among others (Santos-Neto et al., 2019). As a result, for this publication, the following subsection undertakes a literature study on logistic maturity.

Table 3.1 Components of organisational maturity

Type of maturity	Determinants of maturity level
Management and organisational maturity	Shaping the competencies of a leader conditions the effectiveness and efficiency in managing the strategy (flexible adaptation to the needs of the environment), designing the structure, of creating an organisational culture. Great importance is attached to the ability to create a team efficiently achieving the goals set
Praxeological maturity	The ability to correctly define objectives, a high degree of achievement of which enables the optimum satisfaction of stakeholder needs. Effective and efficient actions should result in the required characteristics and properties of products more beneficial than those assumed in the planning phase, in the efficient use of the resources at hand, conditioning at least compliance with the schedule and respecting the assumptions adopted in the budget
Socio-cultural maturity	The determinant of socio-cultural maturity is the ability to create working conditions, interpersonal relations satisfying employees, and relations with the environment satisfying other stakeholders. Key importance should be ascribed to the communication and transfer of information knowledge, which has a significant impact on the ability to develop innovative added value
Technical and technological maturity	Formation of organisational competences conditions the efficient use of new manufacturing methods, the application of innovative technologies leading to the fulfilment of customer requirements with respect to the features and characteristics of the offered product, and the quality of service. The ability to create ergonomic workplaces, securing the required machinery, infrastructure
Ecological maturity	Ability to establish relationships with various interest groups with a real impact on the quality of the organisation's operations, ensuring environmental safety, and meeting legal requirements for environment protection

Source: Łukasiński (2015)

Logistic Maturity

The analysis of the literature allows us to conclude that logistic maturity is an issue that has been taken up by researchers in recent years. The list of literature in this area is not very extensive, but the growth of interest in this research topic is visible. At the same time, one can notice the lack of a comprehensive and complete treatment of the concept of logistic maturity in the conducted analyses. As indicated by Werner-Lewandowska and Kosacka-Olejnik (2020a), in the majority of publications, researchers undertake work on maturity in the context of selected areas of logistics, for example, distribution, supply, and reverse logistics. There have also been studies on supply chain maturity in a general sense, which indicates a holistic approach.

As the beginning of research on the formulation of the concept of logistic maturity, one can assume the years 2012–2013 and three publications published in this period (Battista et al., 2012; Battista & Schiraldi, 2013; Jellouli & Abdelkadhi, 2013). Among Polish researchers, Werner-Lewandowska and Kosacka-Olejnik (2020) have made a significant contribution to the study of logistic maturity in recent years by conducting research on the logistic maturity of service companies. Battista et al. (2012) and Battista and Schiraldi (2013) in their research took action directed at building a logistics process maturity model (LMM—Logistic Maturity Model). Its aim was "to support the enterprises to understand the more critical areas of the process in terms of "immaturity", improvement and the right actions to be undertaken for increasing performances" (Battista et al., 2012). The proposed model was validated in an Italian clothing company. The maturity assessment conducted within the built model took into account the four main logistics areas referring to the SCOR model. The authors did not include reverse logistics processes in their study. These areas were (Battista et al., 2012; Battista & Schiraldi, 2013):

- planning—processes about demand planning,
- procurement—processes about procurement planning, identification and selection of suppliers, and operative management of procurement orders,
- warehousing—processes about stock management, in/out warehouse flow control, storage area management, and goods transportation management,
- distribution—processes about shipment planning and transport management.

The identified processes were subjected to a maturity assessment on a five-point scale. Each level of maturity was linked by the researchers to a set of assumed achievements. The adopted scale is presented in Fig. 3.1.

As mentioned earlier, Jellouli and Abdelkadhi (2013) also conducted a study on maturity in 2012–2013, but the way these authors present this issue in the literature is quite general. Their study was conducted in companies in the industrial region of Gabes, Tunisia. The researchers used a prepared survey questionnaire from which information was collected to assess logistic maturity in terms of four levels:

- Level 0: total absence of Logistics.
- Level 1: non-formalised logistics is reduced to enforcement activities. Considered nonstrategic. Not formalised.

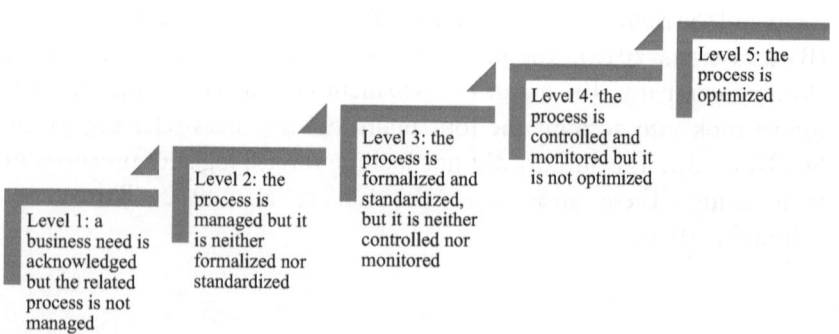

Fig. 3.1 Logistics process maturity levels according to Battista and Schiraldi. Source: compiled from Battista and Schiraldi (2013)

- Level 2: Logistics is fragmented, and partially formalised. It is born between partition functions Branch, Sales, Production, Quality, and Purchasing.
- Level 3: Supply Chain presence of the first signs of a supply chain based on the realisation of the importance of logistics audit to monitor the supply chain.

Werner-Lewandowska and Kosacka-Olejnik (2019a), on the basis of the study by Battista et al. (2012), adopted a definition of logistics maturity as the organisational level of a company indicating to what extent logistics engineering is used in the areas of planning, procurement, warehousing, delivery, and returns. An attempt to study the maturity of logistics processes in the agri-food processing sector, using the Capability Maturity Model Integration (CMMI), was made by Maciejczak (2012). Kalinowski (2012), on the other hand, indicated the possibility of using Business Process Maturity Model (BPMM) to assess the maturity of logistics processes. The author characterises them as sets of recommendations and good practices enabling the achievement of operational effectiveness of implemented logistics processes. In most cases, these models define the current state of the realised processes as a starting point (the so-called as-is approach). On the other hand, their aim is to determine the target state of the realised processes, determining how the logistics processes should be realised (the so-called to-be approach). The target state is often described through the prism of the maturity levels of the implemented processes.

As indicated earlier, extensive research on the logistics maturity of service companies was conducted by Werner-Lewandowska and Kosacka-Olejnik (2018, 2019a, b).

They based their proposed maturity model on three pillars:

- Pillar I—evolutionary phase of industrial logistics—logistics maturity level
- Pillar II—areas of logistics activities of service providers according to the SCOR model,
- Pillar III—logistics engineering tools used in service enterprises.

The main stage of the conducted research was to determine the assumptions of the adopted levels of logistic maturity. The authors assumed that the level of logistic maturity achieved by a company depends on the phase of logistic evolution it is in. Thus, they identified six levels of logistic maturity referring to consecutive phases of industrial logistics development (Table 3.2).

To determine in which phase of the development a company is, the authors measured the use of different types of logistics tools by the studied entities. The list included 81 tools divided into 10 groups: warehouse management tools, transport management tools, inventory management tools, SCM (supply chain management) tools, general management tools, performance management tools, financial management tools and indicators, problem-solving tools, IT tools, and eco-tools. The indicated tools have been assigned to each maturity level. Thus, when measuring the tools used by an enterprise, it is possible to assign it to a given maturity level (Werner-Lewandowska & Kosacka-Olejnik, 2018; Werner-Lewandowska & Kosacka-Olejnik, 2019a, b).

Systematising, the scheme of proceedings in the construction of the logistic maturity model of service enterprises included four essential steps (Fig. 3.2).

It should also be emphasised that the assessment of maturity is not explicit. The assignment to a maturity level applies to each of the logistics areas separately (demand planning, procurement, inventory/storage, distribution, return), assuming that a company can reach different levels of maturity in different areas of logistics activities (Werner-Lewandowska & Kosacka-Olejnik, 2018; Werner-Lewandowska & Kosacka-Olejnik, 2019a, b). In this way, it is possible to create a roadmap that identifies actions to take the company to a higher level of maturity (Werner-Lewandowska & Kosacka-Olejnik, 2020b).

In recent years, publications on research on the maturity of Logistics 4.0 in enterprises have appeared (Oleśków-Szłapka & Stachowiak, 2019; Oleśków-Szłapka et al., 2019; Facchini et al., 2019; Werner-Lewandowska & Kosacka-Olejnik, 2019c). This is a result of the development and dissemination both in practice and in scientific research of the concepts Industry 4.0 and Logistics 4.0. The maturity model of Logistics 4.0 proposed by Oleśków-Szłapka and Stachowiak (2019) included three

Table 3.2 Logistics development phase—level of logistics maturity

Feature	Level 1	Level 2	Level 3	Level 4	Level 5	Level 6
	Fragmentation	Consolidation	Integration	Value added	Globalisation network	Automation
Period of time	To 1960s	To 1980s	1990s	2000s	Twenty-first century	Unknown future
Typical activities	Demand forecasting resource planning procurement/ supplies Storage Stocks of tangible goods manufacture to stock Internal transport Packaging Distribution planning Customer service Transport Other processes	Management of flow of materials physical distribution	Logistic management 3PL BRP	SCM 4PL Environmentalism Sustainable development	Lean SCM SCN GSC E-commerce	Industry 4.0, logistics 4.0 internet of things

Source: Werner-Lewandowska and Kosacka-Olejnik (2018, 2019a, b)

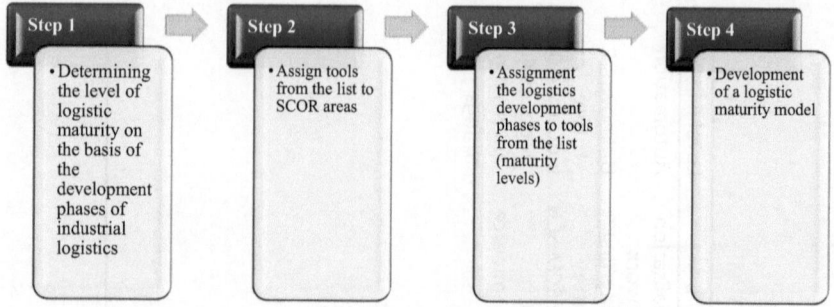

Fig. 3.2 Steps in the logistics maturity model of service companies. Source: compiled from Werner-Lewandowska and Kosacka-Olejnik (2018, 2019a, b)

dimensions: management, flow of materials, and flow of information. According to the authors, the term "Logistics 4.0 maturity" reflects the level to which a company or a supply chain has implemented Logistics 4.0 concepts. The authors distinguish five maturity levels: Ignoring, Defining, Adopting, Managing, and Integrating (Table 3.3).

Werner-Lewandowska and Kosacka-Olejnik (2019c) in their research on the maturity of Logistics 4.0 in enterprises adopted the thesis that the level of maturity is determined by the IT solutions used by the company. Thus, they defined the maturity of Logistics 4.0 as the level of implementation of IT technologies and tools. As the analysis of the literature shows, the issue of logistic maturity is an increasingly popular research problem among scientists. However, it requires further research both in the context of enterprises (of different business profiles) and in cities or regions.

Logistic Maturity of the City

The concept of maturity, especially in process management, as indicated in the previous subsections, is already quite well defined and has been refined theoretically, methodically, and empirically for many years. Research on the maturity of smart cities is not yet advanced.

Cities are evolving towards smart cities and this process is advanced to varying degrees (as indicated in Chap. 1). Therefore, by analysing the

Table 3.3 Logistics 4.0 maturity level

Ignoring	Defining	Adopting	Managing	Integrated
Not aware of the need for integration	See the need for integration but do not know how to manage it	Integration is initiated	Integration at most levels	Full integration resulting in synergy
Do not know about advanced solutions improving material flows	Know about advanced solutions improving material flows but do not use it	Some advanced solutions improving material flows are implemented	Many advanced solutions improving material flows are implemented	All possible advanced solutions improving material flows are implemented
Do not know about advanced solutions improving information flow	Know about advanced solutions improving information flow but do not use it	Some advanced solutions improving information flows are implemented	Many advanced solutions improving information flows are implemented	All possible advanced solutions improving information flows are implemented

Source: Oleśków-Szłapka and Stachowiak (2019)

literature, it is possible to identify emerging models of Smart City maturity. In particular, these are proposed by Maestre-Gongora and Bernal (2019) and Warnecke et al. (2019).

According to the first model, which is largely based on assessing the sophistication of IT solutions in a city (Kettinger & Lee, 2005) and publications identifying criteria for assessing smart cities (Cohen, 2012; Mani & Banerjee, 2015), to assess a city's maturity as a Smart City, Key Domain Area (KDA) should be analysed, such as:

- Strategy—Smart cities have strategies and action plans that identify how investments, including investments in data and digital technologies, enable improved city services and partnerships. An effective strategy focuses on improving outcomes aligned with the city's strategic priorities.

- Data—Smart cities make effective use of their data resources to deliver higher quality services, improve workflows, and make better decisions. Such cities invest in systems that enable data input, integration, and analysis. Open city data provides the basis for transparency, innovation, and engagement.
- Technology—Smart cities are investing in open, flexible, integrated, and scalable ICT architectures that enable effective support of the city's strategic and operational processes, and enable accelerated uptake of innovation and new solutions.
- Administration and service delivery—Smart cities are adapting traditional governance and service delivery models to take advantage of the opportunities associated with the application of technology. They are investing in systematic partnership models that focus on shared outcomes.
- Stakeholder engagement—Stakeholder engagement, in particular city residents, is a key component of a Smart City. Smart cities use a variety of methods to increase residents' involvement in city life. At the same time, they invest in technology and digital data to increase their openness and transparency. Smart cities proactively improve the use of digital services while supporting the digitally excluded.

On this basis, the authors propose the following levels of maturity:

- Capacity Level 0: Capabilities are not made nor exist to achieve the results of the KDA.
- Capacity Level 1: Activities develop incipient capabilities for the domains. Processes have the possibility of some initial results in the KDA.
- Capacity Level 2: Systematisation of processes (orientation, training, and implementation plans) that represent the KDAs. There are some formal evaluation mechanisms and evidence of short-term results.
- Capacity Level 3: Institutionalised and innovative processes, systematic process monitoring, change management, and continuous improvement plans.

A similar model taking into account a broader spectrum of analysis indicates 5 levels of maturity of smart cities (Table 3.4) (Urban Tide, 2016).

Table 3.4 Maturity levels of smart cities

	Level 1	Level 2	Level 3	Level 4	Level 5
City management model	Silo	Cooperation of systems	Integration of systems	Managed system	Sustainable and open system of systems
Status of technology use	Actions aimed at improving services based on digital data	Holistic thinking and the emerging need for data sharing	Digitisation that is strategically led and results-oriented, which takes into account circular technology investments	Technology and data take into account dynamic data updating and ongoing response	A constantly evolving, adaptable peri-urban intelligent system
Impact on results	Focus on recording	Cross-departmental partnership initiatives are established to focus on mutual benefits	Shared responsibility for performance and a joint circular investment programme	Better anticipation, prevention and real-time response.	A peri-urban, open "system of systems" stimulates innovation, which increases the city's competitiveness
Smart City strategy	There is not yet an overall roadmap for digital transformation. Investing only in selected and subjectively isolated areas to ensure record keeping and support of selected tasks within the department/business unit	Strategy exists at the department/ organisational unit level—investments are largely carried out at the department/ organisational unit level—opportunities arise to establish common strategic objectives and joint projects with partners (e.g. other organisational units)	A shared vision, strategy and action plan for a Smart City developed with multiple partners in multiple areas— documented and established business cases and joint investment to deliver scaled improvements and expected shared outcomes	Vision strategy and action plan for a Smart City established at a citywide level—improved service outcomes confirm and support future large-scale service improvements	The strategy is optimised and evolved based on documented impacts on the city's competitiveness— smart investments have a clear impact on the city's strategic priorities

(continued)

Table 3.4 (continued)

	Level 1	Level 2	Level 3	Level 4	Level 5
Information and data	Data reuse and integration are limited by the number of different systems used in different operations. There are issues around data integrity, quality, privacy, and security—data is used primarily to deliver a specific service	Barriers to optimising data resources are discussed between partners and solutions are then developed. Single-unit applications are used for advanced data analysis and sharing—some data sets are made publicly available	A data management and optimisation strategy has been developed and agreed between partners—the city is investing in advanced data management, data collection, analytics, and big data applications—A wide range of open data is published with the strategic intention of using it to drive innovation—citizens create and share data in key areas (e.g. through community data creation initiatives—crowdsourcing)	Data resources are used to provide useful information—enhanced data collection and analysis are used, leading to better decision-making and service design—established open data community creates new services valued by users—citizen willingness to share data is widespread	Data analytics is used to dynamically, automatically predict and preventively improve service delivery—Real-time response to unpredictable events is possible—open data community generating new market opportunities
Technology	ICT system architectures are mainly designed to support specific city tasks—limited investment in application-specific sensor networks	Some shared or integrated ICT architectures exist but are implemented in a limited set of services—barriers to systems integration identified, understood, and resolved between partners. Use cases for sensor networks exist	Investment in the integration of ICT architectures between organisational units is taking place—joint investment plans are being made for the deployment and use of sensor networks across the city	Cross-organisational ICT architectures are already in place. They are scaled and adapted to changing needs—ICT architectures enable the acceleration of innovation in service delivery—a sensor network exists across the city, used for various applications	Business units are continually reviewing, adapting, and investing in ICT architecture to transform and improve service delivery. An efficient network environment is provided throughout the city

Administration and service provision	Leadership, management, and funding focus on service transformation primarily within traditional organisational models—within service delivery, there are traditional customer service provider/commissioner-contractor user relationships and often individual services are managed separately	New governance and leadership models are being tested that encourage collaboration with a wide range of stakeholders (including the private sector) to enable cross-cutting service transformation (across departments and business units)—there is joint budget accountability for some initiatives	Management and leadership models are evolving to enable shared responsibility for delivering system-wide outcomes—greater input into problem-solving and service design from providers, contractors, and users—alignment of budgets and organisational structures to ensure an effective and transparent integrated system-wide approach to service delivery	A transparent and stable multi-stakeholder governance model is in place, ensuring better decision-making and city-wide outcomes—service users have a strong say in how services are delivered—traditional relationships with contractors are evolving to include profit sharing, public-private partnerships, contracting, and success fee accounting	The leadership and management model stimulates an innovation system that promotes new combinations of service delivery and greater effectiveness in influencing the city's strategic priorities
Stakeholder involvement	Stakeholder participation in service improvements is focused on specific services and tasks and is limited by the lack of clear and accessible information on the functioning of city services—opportunities to increase public participation through network and community channels are recognised. Selected initiatives in this area are in place	Investments and projects in digital solutions to engage citizens are being carried out at departmental and unit levels—the focus is on using digital technologies to provide better information, transparency, and stimulate engagement—action is being taken to eliminate digital exclusion in specific areas of service delivery	City-wide and multi-partner strategies have been developed to increase citizen engagement, which includes effective use of digital technologies and digital inclusion—public participation tools, community engagement, and other activities strengthen the voice of stakeholders and citizens	The city uses multiple channels to engage with citizens. Channels are tailored to the needs of specific audiences—the views and ideas of citizens, businesses, and stakeholders are systematically gathered through multiple channels to improve the quality of the city's services—citizens benefit from integrated services in organisations that use the best digital technology for them	The city has established inclusive and personalised engagement models that stimulate innovation and collaborative approaches across the public and private sectors—digital literacy across the population is high and support is provided or alternatives are available for those who need it

Source: Urban Tide (2016)

The research on the maturity of smart cities shows only a slight inclusion of logistics solutions in the city. IT solutions are by far the most strongly emphasised.

The third model is based on combining research on the key differentiators of smart cities (Cohen, 2016) with research on the maturity of IT solutions (Becker et al., 2009), while including another essential factor, namely urban resilience (Desouza & Flanery, 2013). In this model, the authors focus on transport solutions for urban mobility and distinguish components such as policy and planning, ICT integration, intermodal integration, public transport performance, environmental impact, and social impact. In each category, the authors graded the potential solutions in the city by giving a model description of the area and the factors analysed (Table 3.5).

The third model has a strong focus on urban mobility. Taking into account the research results from Sect. 3.2, it should be complemented with broader logistical aspects. Such a research direction was proposed by Kiba-Janiak (2018), followed by Kijewska et al. (2021). The proposals concern the assessment of the level of maturity in terms of the involvement of freight transport stakeholders in the city (Kijewska et al., 2021), maturity in terms of the formulation and implementation of logistics strategies in the city (transport aspects in city strategies are included) (Kiba-Janiak, 2018), freight transport aspects in city strategic planning (Kiba-Janiak, 2021) and integrated and sustainable passenger and freight transport strategies in the city (Kiba-Janiak et al., 2021). Thus, on the one hand, it is noteworthy that researchers refer to the category of maturity, including logistical aspects, to cities. On the other hand, however, it should be emphasised that this research does not exhaust the problem of formulating and studying the logistic maturity category of cities, especially in combination with the Smart City issue. In the study of the city's maturity in the formulation and implementation of logistics strategies that take into account aspects of the transport of people and goods in the city, researchers use two main dimensions (Kiba-Janiak, 2018, 2021): the degree of advancement in the formulation of the city's logistics strategy and the extent of implementation of the city's logistics strategy. Each dimension is characterised by a number of identified criteria. On this basis, five levels of city maturity were determined (Fig. 3.3).

Table 3.5 Smart city areas with recognised roles for transport

Policy and planning	Relevance	The urban mobility strategy provides guiding principles pertaining to development, investment, and policy-making decisions in the long term. It should be designed to promote a more efficient, safe, environmentally friendly, and equitable mobility system based on the city's individual conditions. Ideally, the mobility strategy is to be developed in a collaborative process involving all stakeholders, such as city and regional authorities, local businesses, and citizens. It should integrate urban transport data and include strategies for emergency and disaster management and for mitigating the effects of ecological and social disruptions
	Indicators	Existence of long-term mobility vision, focus of mobility strategy, existence of surveillance and response system, focus of resilience measures, existence of traffic flow monitoring, existence of traffic flow forecasting
ICT integration	Relevance	Smart mobility enables a more efficient, safe, environmentally friendly, and equitable mobility system through the integration of ICT. Innovative ICTS can provide the basis of new mobility concepts such as car sharing and intermodal transport provision and should facilitate the optimal utilisation of existing data and infrastructure
	Indicators	Focus of intelligent traffic management system, percentage of public vehicles equipped with automatic vehicle location (AVL), provision of intelligent parking guidance system, provision of real-time public transport information, use of automatic number plate recognition (ANPR) for traffic control and law enforcement use of dynamic route guidance for public/emergency vehicles existence of early emergency/failure detection system level of smart card use

(continued)

Table 3.5 (continued)

Intermodal integration	Relevance	Intermodal integration, i.e., the combination of different modes of transport, represents a well-recognised strategy for reducing the environmental impact of transport and relieving the strain on the road network. In an optimised smart mobility system, each mode of transport should be used according to its unique benefits and an integrated solution combining various modes within a transportation chain should be sought
	Indicators	Share of public transport, walking, and cycling in modal split public transport network density, cycle lane density, car sharing stations per 1000 inhabitants, bike sharing stations per 1000 inhabitants, intermodal fare integration vehicles registered per 1000 inhabitants
Public transport performance	Relevance	To generate the highest possible value for citizens and other stakeholders, the transport system's performance should be monitored and evaluated continuously. Based on the city's individual conditions, requirements, and evaluation criteria, strategies for improvement and adaptation should be initiated
	Indicators	Mean travel, time to work; access to public transport within walking distance; satisfaction with public transport, percentage of the system experiencing congestion hours of congestion
Environmental impact	Relevance	Smart mobility concepts have the potential to significantly improve the environmental impact of urban transport by reducing pollution, greenhouse gas emissions, and noise and contributing to a more pleasant and healthier urban environment. Reduction of environmental impact should be a major guiding principle in planning and implementing urban mobility strategies
	Indicators	Transport-related CO_2 emissions, mean PM10 concentration, mean $_2$ concentration, share of renewable energy carriers in public transport system, noise pollution in residential areas

(continued)

Table 3.5 (continued)

Social impact	Relevance	Urban transport provides access to numerous human activities, including means of subsistence, medical care, education, and social activities. For this reason, the urban mobility system must be aligned to acknowledge the needs of all societal groups and members of society. It should be designed to promote a more efficient, safe, environmentally friendly, and equitable urban transport system based on the city's individual conditions
	Indicators	Number of transport-related fatalities, number of traffic incidents, opportunity for active mobility, affordability index of public transport for the poorest quartile of population, accessibility of public transport for handicapped persons

Source: Desouza and Flanery (2013)

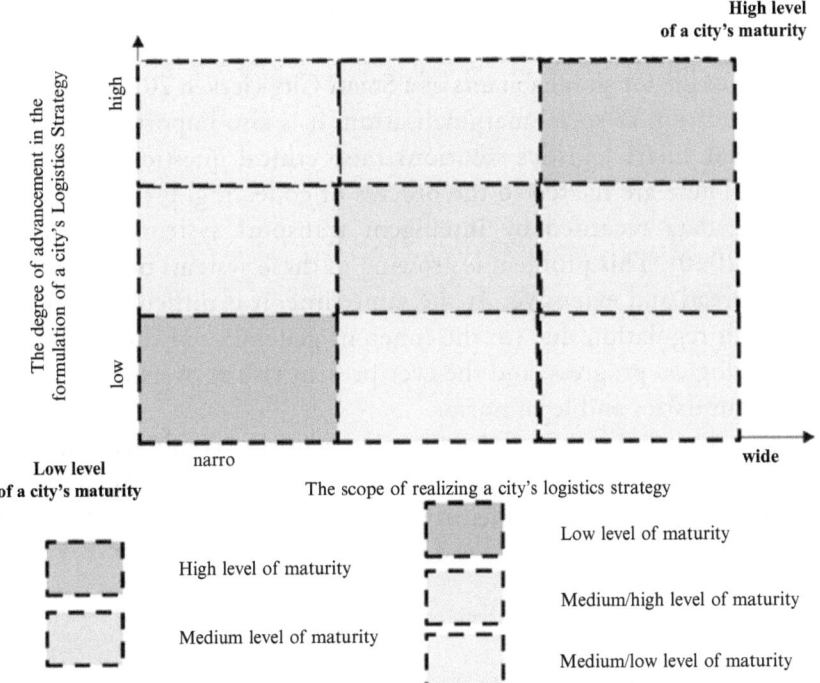

Fig. 3.3 City logistic maturity model for strategy formulation and implementation. Source: Kiba-Janiak et al. (2021)

Taking into account the body of literature in the area of Smart City development level, as well as city logistics maturity levels, it is worth directing the research towards those aspects that combine these two constructs. As indicated, the logistics component in Smart City is focused on the use of state-of-the-art IT and ICT technologies in technical urban systems (Bruska, 2012). In terms of outcomes, it is admittedly accompanied by economic issues related to cost reduction and social issues related to improved safety and convenience (Navarro et al., 2016). Nevertheless, it is important not to forget the broader, holistic view of Smart City logistics issues, which in an over-technicised form can lead to social exclusion, including in particular the digital exclusion of older people or residents with disabilities (Porru et al., 2020; Zhang et al., 2020; Mora et al., 2019; Jittrapirom et al., 2019).

For these reasons, smart logistics solutions should be subject to a social dialogue that takes into account the needs of those who, for various reasons, may be overlooked. Expectations identified in this way should then be used in the design, testing, and implementation of urban logistics systems, as this is a prerequisite for sustainable development and a high quality of life for all inhabitants of a Smart City (Jeekel, 2017).

In addition to social marginalisation, it is also important to mention that smart logistics solutions raise ethical questions and concerns. These are related to the process of collecting, processing, and storing data recorded by intelligent transport systems (Gutiérrez et al., 2020). This problem is growing as these systems become more widespread and extensive. At the same time, it is difficult to solve it through regulation due to: the inherent generality of the law, rapid technological progress, and the ever-present risk of overstepping ethical boundaries and legal norms.

The identified levels of city logistics maturity, combined with the level of development of smart cities, allow cities to be compared with each other and possibly identify actions to achieve higher levels of maturity.

References

Albliwi, S. A., Antony, J., & Arshed, N. (2014). Critical literature review on maturity models for business process excellence. *IEEE international conference on industrial engineering and engineering management, 9–12 Dec* (pp. 79–83).

Battista, C., Fumi, A., & Schiraldi, M. M. (2012). The logistic maturity model: Guidelines for logistic processes continuous improvement In *Proceedings of the POMS 23rd annual conference, Chicago, IL, USA* (pp. 1–18).

Battista, C., & Schiraldi, M. M. (2013). The logistic maturity model: Application to a fashion company. *International Journal of Engineering Business Management, 5*(21), 1–11.

Becker, J., Knackstedt, R., & Pöppelbuß, J. (2009). Developing maturity models for IT management. *Business and Information Systems Engineering, 1*(3), 213–222.

Bruska, A. (2012). Logistyka jako komponent smart city. *Studia Miejskie, 6.*

Cohen, B. (2012). The smartest cities in the world 2015: Methodology. Retrieved January 4, 2017 from https://www.fastcompany.com/3038818/the-smartest-cities-in-the-world-2015-methodology.

Cohen, B. (2016). 6 components for smart cities, smart cities wheel. UBM's future cities. Available at: www.ubmfuturecities.com/author.asp?section_id=219&doc_id=524053.

Crosby, P. (1979). *Quality is free.* McGraw-Hill.

de Boer, F. G., Müller, C. J., & ten Caten, C. S. (2015). Assessment model for organizational business process maturity with a focus on BPM governance practices. *Business Process Management Journal, 21*(4), 908–927.

Desouza, K. C., & Flanery, T. H. (2013). Designing, planning, and managing resilient cities: A conceptual framework. *Cities, 35*, 88–89.

Dijkman, R., Lammers, S. V., & de Jong, A. (2016). Properties that influence business process management maturity and its effect on organizational performance. *Information Systems Frontiers, 18*(4), 717–734.

Facchini, F., Oleśków-Szłapka, J., Ranieri, L., & Urbinati, A. (2019). A maturity model for logistics 4.0: An empirical analysis and a roadmap for future research. *Sustainability, 12*(86), 1–18.

Gutiérrez, A., Domènech, A., Zaragozí, B., & Miravet, D. (2020). Profiling tourists' use of public transport through smart travel card data. *Journal of Transport Geography.* 88102820-S0966692320302283 102820. https://doi.org/10.1016/j.jtrangeo.2020.102820

Harmon, P. (2014). Evaluating an organization's business process maturity. *Business Process Trends, 2*(3), 1–11.

Hüner, K. M., Ofner, M., & Otto, B. (2009). Towards a maturity model for corporate data quality management, In: *Proceedings of the 2009 ACM symposium on applied computing—SAC '09* (p. 231).

Jeekel, H. (2017). Social Sustainability and Smart Mobility: Exploring the relationship. *Transportation Research Procedia, 25*, 4296–4310.

Jellouli, O., & Abdelkadhi, M. (2013). Test logistics maturity of the industrial zone in the region of Gabes. *International Journal of Supply Chain Management, 2*(4), 71–75.

Jittrapirom, P., van Neerven, W., Martens, K., Trampe, D., & Meurs, H. (2019). The Dutch elderly's preferences toward a smart demand-responsive transport service. *Research in Transportation Business & Management, 30*, 100383.

Kalinowski, T. B. (2012). Koncepcja oceny dojrzałości procesów logistycznych. *Zeszyty Naukowe, Uniwersytet Ekonomiczny w Poznaniu, 224*, 37–47.

Kenny, J. (2006). Strategy and the learning organization: A maturity model for the formation of strategy. *The Learning Organization, 13*(4), 353–368. https://doi.org/10.1108/09696470610667733

Kerzner, H. (2001). *Strategic planning for project management using a project management maturity model* (pp. 2–235). Wiley.

Kettinger, W. J., & Lee, C. C. (2005). Zones of tolerance: Alternative scales for measuring information systems service quality. *MIS Quarterly & The Society for Information Management., 29*(4), 607–623.

Kiba-Janiak, M. (2018). *Logistyka w strategiach rozwoju miast*. Wydawnictwo Uniwersytetu Ekonomicznego we Wrocławiu.

Kiba-Janiak, M. (2021). Urban freight transport in city strategic planning. *Research in Transportation Business & Management*, 1–13.

Kiba-Janiak, M., Thompson, R., & Cheba, K. (2021). An assessment tool of the formulation and implementation a sustainable integrated passenger and freight transport strategies. An example of selected European and Australian cities. *Sustainable Cities and Society, 71*, 1–10.

Kijewska, K., de Oliveira, L., dos Santos, O., Bertoncini, B., Iwan, S., & Eidhammer, O. (2021). Proposing a tool for assessing the level of maturity for the engagement of urban freight transport stakeholders: A comparison between Brazil, Norway, and Poland. *Sustainable Cities and Society, 72*, 1–10.

Kosieradzka, A. (2016). Modele dojrzałości jako narzędzie stymulowania zrównoważonego rozwoju organizacji. In J. Ejdys (Ed.), *Społeczna odpowiedzialność i zrównoważony rozwój w naukach o zarządzaniu.* TNOiK Dom Organizatora.

Kosieradzka, A., & Smagowicz, J. (2017). *Analiza porównawcza modeli dojrzałości organizacji* (pp. 291–292). UE. http://koncepcje.uek.krakow.pl/wp-content/uploads/2017/01/20_2016.pdf

Looy, A. (2014). *Business process maturity: A comparative study on a sample of business process maturity model.* Springer.

Łukasiński, W. (2015). Dojrzałość jakościowa organizacji na przykładzie działu kruszyw. *Przegląd Organizacji, 5*(904), 30–36.

Maciejczak, M. (2012). Dojrzałość procesów logistycznych w sektorze przetwórstwa rolno-spożywczego wg modelu CMMI. *Logistyka, 6,* 513–518.

Maestre-Gongora, G., & Bernal, W. (2019). Conceptual model of information technology management for Smart Cities: SmarTICity. *Journal of Global Information Management, 27*(2).

Mani, D., Banerjee, S. (2015). Smart City maturity model (SCMM)-BSI. Retrieved May 14, 2017 from http://isbinsight.isb.edu/smart-city-maturity-model-scmm.

Mora, L., Deakin, M., & Reid, A. (2019). Strategic principles for smart city development: A multiple case study analysis of European best practices. *Technological Forecasting & Social Change, 142,* 70–97.

Mutafelija, B., & Stromberg, H. (2008). Capability maturity model integration (CMMI), In *Process improvement with CMMI® v1.2 and ISO standards* (pp. 25–63).

Navarro, C., Roca-Riu, M., Furió, S., & Estrada, M. (2016). Designing new models for energy efficiency in urban freight transport for smart cities and its application to the Spanish case. *Transportation Research Procedia, 12,* 314–324.

Nolan, R. (1973). Managing the crisis in data processing. *Harvard Business Review, 57*(2), 115–126.

OGC. (2007). Zarządzanie ryzykiem: przewodnik dla praktyków.

Oleśków-Szłapka, J., & Stachowiak, A. (2019). The framework of logistics 4.0 maturity model. In A. Burduk, E. Chlebus, T. Nowakowski, & A. Tubis (Eds.), *Intelligent systems in production engineering and maintenance. ISPEM 2018* (pp. 1–11). Springer.

Oleśków-Szłapka, J., Wojciechowski, H., Domański, R., & Pawłowski, G. (2019). Logistics 4.0 maturity levels assessed based on GDM (Grey decision model) and artificial intelligence in logistics 4.0—trends and future perspective. *Procedia Manufacturing, 39,* 1734–1742.

Pöppelbuß, J., Röglinger, M. (2011). What makes a useful maturity model? A framework of general design principles for maturity models and its demonstration in business process management., European Conference on Information Systems Proceedings.

Porru, S., Misso, F. E., Pani, F. E., & Repetto, C. (2020). Smart mobility and public transport: Opportunities and challenges in rural and urban areas. *Journal of Traffic and Transportation Engineering, 7*(1), 88–97.

Rohloff, M. (2009). Case study and maturity model for business process management implementation. *Lecture Notes in Computer Science, 57*(1), 128–142.

Rosemann, M., & de Bruin, T. (2005). Application of a holistic model for determining BPM maturity. *BP Trends.* http://bpm-training.com/wpcontent/uploads/2010/04/applicationholistic

Santos-Neto, J. B. S., Costa, A., & P. (2019). Enterprise maturity models: A systematic literature review. *Enterprise Information Systems, 13*(5), 719–769. https://doi.org/10.1080/17517575.2019.1575986

Skrzypek, E. (2013). Uwarunkowania i konsekwencje osiągania dojrzałości organizacyjnej w warunkach zmienności otoczenia. In E. Skrzypczyk (Ed.), *Dojrzałość organizacji–aspekty jakościowe* (p. 36). Wydawnictwo UMCS.

Tarhan, A., Turetken, O., & Reijers, H. A. (2016). *Business process maturity models: A systematic literature review* (Vol. 75, pp. 122–134). Information and Software Technology.

Urban Tide, Overview of the smart cities maturity model (2016). https://static1.squarespace.com/static/5527ba84e4b09a3d0e89e14d/t/55aebffce4b0f8960472ef49/1437515772651/UT_Smart_Model_FINAL.pdf.

Warnecke, D., Wittstock, R., & Teuteberg, F. (2019). Benchmarking of European smart cities—a maturity model and web-based self-assessment tool. *Sustainability Accounting, Management and Policy Journal, 10*(4), 654–684.

Werner-Lewandowska, K., & Kosacka-Olejnik, M. (2018). Logistics maturity model for service company—theoretical background. *Procedia Manufacturing, 17*, 791–802.

Werner-Lewandowska, K., & Kosacka-Olejnik, M. (2019a). Model dojrzałości logistycznej przedsiębiorstw usługowych—podstawy teoretyczne. In A. Bujak, A. Gębczyńska, & M. Brzozowska (Eds.), *Przedsiębiorczość i Zarządzanie. Logistyka w naukach o zarządzaniu* (Vol. 7, pp. 175–189). Wydawnictwo Społecznej Akademii Nauk, , tom XX.

Werner-Lewandowska, K., & Kosacka-Olejnik, M. (2019b). Logistics maturity model for engineering management—method proposal. *Management Systems in Production Engineering, 27*(1), 33–39.

Werner-Lewandowska, K., & Kosacka-Olejnik, M. (2019c). Logistics 4.0 maturity in service industry: Empirical research results. *Proceedia Manufacturing, 38*, 1058–1065.

Werner-Lewandowska, K., & Kosacka-Olejnik, M. (2020a). *Dojrzałość logistyczna przedsiębiorstw usługowych.* Instytut Naukowo Wydawniczy Spatium.

Werner-Lewandowska, K., & Kosacka-Olejnik, M. (2020b). How to improve logistics maturity ?—a roadmap proposal for the service industry. *Procedia Manufacturing, 51*, 1650–1656.

Wibowo, M. A., & Waluyo, R. (2015). Knowledge management maturity in construction companies. *Procedia Engineering, 125*, 89–94.

Wolniak, R. (2019). The level of maturity of quality management systems in Poland—results of empirical research. *Sustainability, 11*, 4239.

Zhang, M., Zhao, P., & Qiauo, S. (2020). Smartness-induced transport inequality: Privacy concern, lacking knowledge of smartphone use and unequal access to transport information. *Transport Policy, 99*, 175–185.

4

Logistics Innovation in Smart Cities

Innovation in City Logistics

Innovation is one of the key areas of interest for both researchers and management practitioners. For a number of years, publications on innovation mainly referred to enterprises of different sizes or types of activities in the market. However, recent years have brought increasing interest in innovation in the city, including in-city logistics. Mainly, this is the result of an increasing focus by researchers and city managers on the pursuit of improving the quality of life in cities, on the concept of sustainable urban development, and consequently on the creation of Smart City. Jin et al. (2021) point out that innovation drives urban development, including logistical aspects, and determines whether cities will be able to achieve sustainability. Karvonen et al. (2014) emphasise the need for both public and private urban actors to seek innovative solutions as a result of increasing complexity and growing urban problems. The literature also highlights the links between innovation and the Smart City concept (Nilssen, 2019; Nam & Pardo, 2011). Szołtysek (2016) identifies the innovative city with the city on the road to the Smart City. Nam and Pardo (2011) presented a framework for Smart City innovation with three dimensions of innovation elaborated: technology, organisation, and policy. While Nilssen (2019) points to four dimensions of Smart City innovation: the technological dimension, the organisational dimension, the collaborative dimension, and the experimental dimension. Innovations in city logistics represent an opportunity for local governments, as they allow building the position of a city, and shaping its image in a modern way, which stands out from other cities. Moreover, they provide a clear signal to investors, indicating the open and entrepreneurial character of the city (Herbuś, 2015). As Rodríguez-Pose et al. (2021) point out, innovations are more often concentrated in larger and more economically developed cities and regions. This is mainly due to the resources available. Less developed cities have far fewer resources available for innovation activities. The shortage of human, physical, and financial capital definitely weakens both the generation and the application of innovative solutions. Some of the less developed cities are also geographically limited. Peripheral location causes that the city lies outside the boundaries of diffusion of knowledge coming from more innovative territories and cities. Thus, opportunities to complement locally generated knowledge with nonlocal knowledge are limited (Moreno et al., 2005; Rodríguez-Pose & Crescenzi, 2008; Sonn & Storper, 2008). Significant problems in the implementation of urban innovation are often highly formalised internal procedures concerning, for example, the execution of contracts, the allocation of resources, or the issuing of building permits. Unfortunately, these often cause delays, thus there is a need for reform in urban decision-making processes, the introduction of ad hoc procedures that support the development of innovation, and strong and innovation-friendly local leadership (Aparicio, 2020). Moreover, the lack of public-private cooperation in the context of the triple helix (research institutions, companies, and local authorities) is a significant limiting factor for the development of innovation in city logistics (Verlinde & Macharis, 2016). The concept of innovation has been widely discussed and analysed in the literature, starting with Schumpeter's most well-known and widespread definition, which was introduced to the economic sciences in the 1960s. The definition of innovation in cities is based on the developed, well-known approaches to innovation in enterprises. Szromnik (2012) points out that an innovative city is such a settlement unit with the status of a city, whose inhabitants, institutions, entrepreneurs, and companies are consistently and systematically oriented towards the creation of knowledge, thanks to which, in cooperation with the local government and R&D institutions, they create conditions for shaping new ideas, concepts, organisational solutions, technologies, and products themselves, better meeting the expectations of all entities of the city's socio-urban system. At the same time, it should be emphasised that the city is innovative not by definition, but as a result of the long-term implementation of partial strategies, evolution, and long-term process of change enabling to reach the idea of an innovative city (Fig. 4.1).

The implementation of typical logistics processes for city development necessitates the implementation of innovations in them that will contribute to a high quality of life in the city, its sustainable development, and high competitiveness (Nowicka, 2015). Innovations in city logistics can be considered as new activities and solutions, even if they are of a secondary nature, as long as they are new to the local stakeholders or the governing structure of a given city, which emerged as a result of deliberate actions and contributed to the progress of the city's development in social, economic, and environmental dimensions (Pluta-Zaremba, 2015). The innovations implemented are intended to bring a number of tangible benefits. However, the issue of a clear, reliable evaluation, applicable to different categories of innovation, remains a problem. To this end, Patier and Browne (2010) propose an evaluation methodology that consists of 60 different metrics and indicators that allow full consideration of social, economic, and environmental impacts.

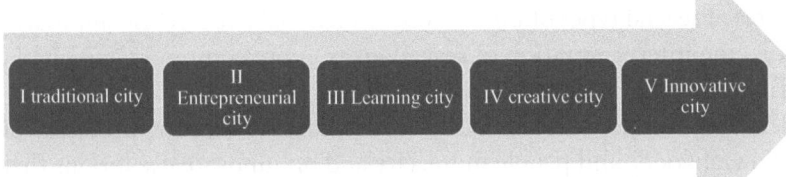

Fig. 4.1 Evolution towards an innovative city. Source: compiled from Szromnik (2012)

Table 4.1 Examples of types of innovation in city logistics

Type of innovation	Interpretation
Product	Introducing innovative products within city logistics—car sharing, carpooling, city terminals, new fare systems
Process	Improving transport system management activities, logistics processes, technical equipment, modern technologies in public transport, freight, increased efficiency of operations, better reliability results
Organisational	Modification of the structures of the entities responsible for the organisation and implementation of the city logistics concept (both on the municipal side and on economic units), e.g., change of traffic organisation, introduction of bus lanes, inner-city traffic-calmed zones
Marketing	Marketing activities, city marketing, social marketing, promotional activities, information campaigns on implemented solutions

Source: Niedzielski and Tundys (2012)

As mentioned earlier, the whole nomenclature and taxonomy related to innovation have been developed in the context of industrial innovation. Among the most basic classifications is the division into product, process, organisational, and marketing innovations. This classification can be successfully adapted to innovations in city logistics. Table 4.1 shows examples of innovation in city logistics taking into account the four categories. However, it should be remembered that in the case of many innovative solutions, it is difficult to talk about their assignment to only one category. The distinction between different types of innovations is not unambiguous, which results from the inhomogeneous nature of the introduced changes. Often, the implemented solutions combine the features of several types of innovations, which make the hybrid approach, that is, the interpenetration of several areas, quite common (Niedzielski & Tundys, 2012; Przybylska et al., 2016).

The literature especially focuses a lot of attention on the problems of the flow of goods and people in the city and on innovative solutions that reduce them and fit into the policy of sustainable development (both environmental, social and economic pillars). In their research, Cichosz et al. (2014) ask the question: Is the process of balancing urban transport a catalyst for urban innovation in this area? Innovation is understood here as the willingness and potential of a city to implement transport innovations, that is, innovations that improve the efficiency and effectiveness of a place for a given community (Szelągowska, 2014), but above all as a result of already implemented innovative projects in this area (Calantone et al., 2002).

A large part of the innovative solutions are related to passenger transport, both in relation to individual and collective transport. These solutions are mainly based on the aim to reduce the movement of road transport modes in the city through (Kiba-Janiak, 2012; Szołtysek, 2016; Jaroszyński & Chłąd, 2015):

- development of micromobility to enable multimodal travel (urban bicycle system, urban scooter system, urban moped system),
- introducing charges for entering city centres, with possible exemptions for alternative propulsion vehicles or vehicles with at least nine seats,
- total or partial restrictions on cars entering city centres (at certain times of the year, week, or days),
- development of carpooling, car sharing,
- introducing incentives to use public transport.

The use of public transport is indicated as a major factor in reducing road transport in the city and thus reducing external transport costs and balancing the flow of people in the city. A number of innovative solutions implemented in cities are supposed to contribute to this: city card system (Grad et al., 2013; Jaroszyński & Chłąd, 2015), dynamic (real) passenger information system (Grad et al., 2013; Kiba-Janiak, 2012), systems and services for mobile information, organisation, and integration of public transport services, land use planning and traffic management (infomobility) (Bellandi et al., 2014), preferences for public transport, for example, dedicated bus lanes (Kiba-Janiak, 2012), introduction of small buses moving at a higher frequency than buses (Kiba-Janiak, 2012). Other innovative solutions implemented in cities with reference to passenger transport include (Kiba-Janiak, 2012; Jaroszyński & Chłąd, 2015; Iwan et al., 2018) transport telematics (mainly intelligent transport systems, adaptive traffic light control systems), development of ecological branches and means of transport, for example, electric or low-emission vehicles, systems aiming to develop multimodal transport, for example, Park&Ride, Bike&Ride.

A separate category is an innovation in freight transport, which has been given increasing importance in recent years for urban development. The main innovative solutions implemented in cities include:

- introducing charges for entering city centres (similar to passenger transport) (Kiba-Janiak, 2012),
- total or partial restriction of access to cities for heavy goods vehicles. These restrictions may apply to all heavy goods vehicles or only those exceeding designated technical parameters (weight, dimensions) as well as emission standards (low-emission zones in the city). They may refer to dedicated areas in the city and designated hours or days (Kiba-Janiak, 2012; Taniguchi et al., 2014),
- setting delivery times for trucks (off-peak hours) (Kiba-Janiak, 2012),
- organisation of night-time deliveries to retail, service, or production outlets (Kiba-Janiak, 2012),
- creation of logistics entities, such as micro-hubs, integrated logistics centres, urban consolidation centres, and urban distribution centres. They fit into supply chains by organising and coordinating last-mile delivery and reverse logistics. They can also constitute distribution multimodal facilities. They are a form of logistics objects of warehouse type with buffer, distribution, and completion specificity. Moreover, noteworthy is the idea of sharing the mentioned facilities by several logistics entities. (Heitz & Beziat, 2016; Montwiłł et al., 2021; Nathnail et al., 2016; Fijałkowski, 2010; Russo et al., 2021; Kauf, 2016),
- use of electric and low-emission vans, freight trams, delivery bicycles, as well as other alternative-powered vehicles together with rational transport fleet management (Taniguchi et al., 2014, 2020; Bellandi et al., 2014; Iwan et al., 2018; Kauf, 2016),
- optimisation of daily goods deliveries by planning vehicle routes on the assumption of less fuel consumption, circling around the city, and consequently less exhaust emissions (Taniguchi et al., 2014, 2020; Bellandi et al., 2014),

- use of telematics technologies including ITS (Jaroszyński & Chłąd, 2015; Kozerska & Konopka, 2018; Koźlak, 2008; Kręt, 2020; Selwon & Roman, 2017; Wojewódzka-Król & Rolbiecki, 2010; Taniguchi et al., 2020),
- construction of underground transport systems, for example, CargoCap (Kauf, 2016),
- construction of tube systems (Hyperloop) with small compartments or single palette transport capacity (Müller et al., 2019),
- Development of autonomous vehicles for last-mile delivery (Taniguchi et al., 2020).

A major challenge in the area of urban freight transport is parcel delivery associated with strongly growing e-commerce. These deliveries are a specific segment of freight transport, mainly due to their time constraints for delivery (fast delivery to recipients). In addition, they require a network of transhipment terminals, sending and receiving points, and a number of transport operators. Yet another issue remains, the return of purchased goods, the scale of which is steadily increasing.

E-commerce generates a very large scale of highly fragmented deliveries of consignments to diverse and highly dispersed customers. This results in very complex logistical needs in a very limited urban environment. In this situation, both logistics companies and city managers face a large challenge to organise sustainable shipment flows (Heitz & Beziat, 2016; Montwiłł et al., 2021; Nathnail et al., 2016). Both already implemented and tested modern innovative solutions come to the rescue. Examples include development of alternative delivery and collection points based on locations other than the customer's home (e.g. the nearest post office, shop, or petrol station) (Vural & Aktepe, 2021); development of parcel machines (Taniguchi et al., 2020); use of drones in parcel delivery (Müller et al., 2019); the development of urban consolidation or distribution centres; the use of alternative means of transport (e.g. delivery bicycles); the development of the crowdshipping concept referred to as crowd logistics (based on the idea of crowdsourcing, which involves the use of the so-called crowd potential)—that is, the involvement in the delivery of parcels of those people who move around the city for other purposes. (Taniguchi et al., 2020; Pan et al., 2015; Serafini et al., 2018; Chen & Pan, 2015). Viu-Roig and Alvarez-Palau (2020) divide innovations in last-mile e-commerce delivery into three groups: organisational, technology-enabled, and data-technique-enabled (Fig. 4.2).

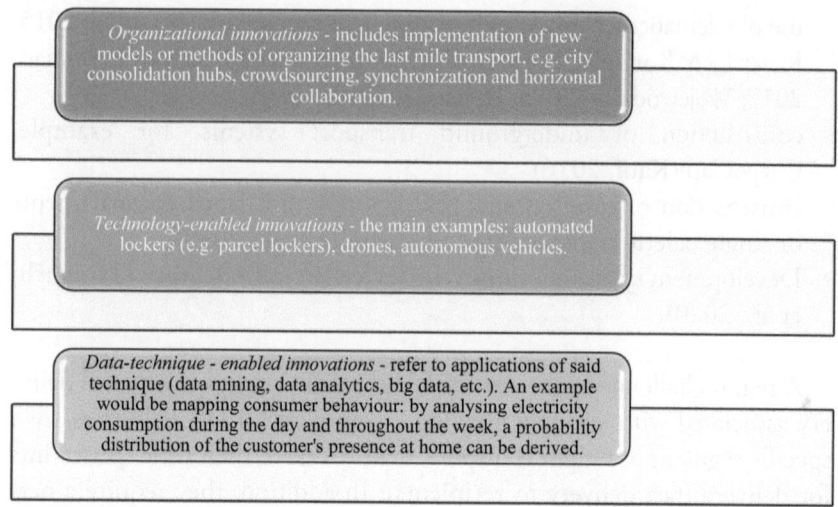

Fig. 4.2 Types of innovation in e-commerce last-mile delivery. Source: compiled from Viu-Roig and Alvarez-Palau (2020)

The selected innovations, indicated, in the context of passenger and freight transport, fit well with the concept of eco-innovation and the sharing economy. Eco-innovation is a key element of environmental policy, protecting the environment and fully achieving sustainable development. The term sharing economy was first coined in 2008 and means sharing, exchanging, and renting out unused resources; otherwise it is defined as joint consumption. The main benefits achievable through innovative solutions that fit into the sharing economy in city logistics are higher efficiency, lower costs, less congestion, lower CO_2, better use of resources, and coping with skills shortages (Van Duin et al., 2020; Digiesi et al., 2017; Pacheco et al., 2018; Standing et al., 2018). The idea of a sharing economy has a much broader application in relation to innovation in city logistics. Other examples are also (Szołtysek, 2016) a system of bicycles, scooters, city mopeds, the idea of car sharing, carpooling, or crowdsourcing.

The last area of innovation in city logistics concerns waste management. This issue has become essential over the years mainly due to the change in approach and the move away from the typical (linear) waste management model towards a closed (circular) economy. The new model is supported by innovations in the area of logistics concerning the collection, sorting, and transport of waste in the city. The basis of the created and implemented modern solutions in the area of waste logistics is the use of information technology (IT), information and communication technology (ICT), machine learning algorithms, and the idea of Internet of Things (IoT) (Sinha et al., 2021; Marques et al., 2019; Ashwin et al., 2021; Nakandhrakumar et al., 2021; Lewicki, 2018). The main innovation solutions in the area of waste logistics are concentrated in four areas:

- IT systems dedicated to various participants in waste management (entities responsible for managing waste management in the city, waste treatment plants, entities transporting waste). To obtain their full functionality, they cooperate with solutions such as barcodes, QR codes, RFID systems, GPS systems, etc. (Jąderko et al., 2016; Ferrer & Alba, 2019),
- IT systems or mobile applications dedicated to residents or property owners (Jąderko et al., 2016),
- robotisation in the design of machines to support waste segregation processes (e.g. hazardous waste) (Karnan et al., 2021; Nakandhrakumar et al., 2021),
- Smart waste bins—supported by ICTs—allow for fully effective and rational waste collection and its full monitoring. Additionally, they can eliminate waste nuisance related to its appearance or smell (Sinha et al., 2021; Rekha et al., 2021; Mamun et al., 2016).

All the innovative solutions applied in waste logistics make it possible to better implement and control selective waste collection and transport processes. They make it possible to rationalise the frequency of waste collection, control the level of filling containers, and thus make it possible to optimise transport operations. Consequently, it is possible to speak of the optimisation of processes in economic, ecological, and social terms, which is part of meeting the objectives of sustainable development (Ferrer & Alba, 2019; Rekha et al., 2021).

In conclusion, the high level of scientific interest in the issue of innovation in city logistics and the comprehensiveness of the discussed topic should be emphasised. The creation and the implementation of innovative solutions in the era of sustainable development have become a necessity in cities. However, it should not be forgotten that this is a major technological, economic, organisational, social, and administrative challenge for city managers. This is all the more reason why it requires close cooperation between all parties involved.

Intelligent Logistics Solutions in Cities: Case Studies

In practice, Smart City logistics solutions take the form of concrete initiatives, projects, and implementations located in different areas and are characterised by varying degrees of technological sophistication (Ydersbonda et al., 2020). This subsection presents six case studies developed on the example of three European cities and three Polish cities. These are cities that stand out in terms of the number and scope of smart logistics implementations and are very often mentioned in international Smart City rankings. Their actions are examples of best practice and contribute to the dynamic development of smart cities in Poland and worldwide.

In the international context, the cities selected for analysis were Amsterdam (The Netherlands), Barcelona (Spain), and Copenhagen (Denmark). A description of Smart City logistics development activities for the former European cities is presented in Table 4.2.

The scale of activities carried out by the Amsterdam city authorities in the area of Smart City logistics is impressive, both in terms of the number of initiatives taken and their scope (Hirschhorna et al., 2019). Their undoubted advantage is their comprehensiveness, manifested primarily in the modernisation and improvement of virtually all forms of transport. Attention is also drawn to the pursuit of maximum greening of city logistics in line with the concept of a city that is not only intelligent but also sustainable. It is also worth noting the design, implementation, and use

Table 4.2 Smart city logistics measures in Amsterdam (The Netherlands)

Description of activities	Results of actions
Greening the transport	Bicycle facilities (about one-third of road transport is done by bicycle)
	Significant increase in the number of electric cars
	Significant growth in car sharing services operated on the central together platform
Improving urban freight flows	Consolidating loads in distribution centres and transporting them using electric vehicles
	Launch of a subsidy scheme for environmentally friendly means of transport of goods
	Support for overnight deliveries
	Designation of routes for freight traffic
	Use of electrically powered boats for freight transport
	Creating the world's most sustainable port in a Ship-to-grid project
Implementation of projects and applications facilitating transport for residents	Smart Parking: smart parking system using a ticket that never expires
	KeyPass: mobile automatic unmanned ticket purchasing system
	MobilityLabel: an application to help you find the best location for your business from a mobility perspective
	Creation of an urban platform as a place of communication between authorities, private entities, research institutions, and citizens in the field of initiating Smart City solutions, including in particular the reduction of carbon emissions
Partnership for the development of smart mobility	Partnership for the development of smart mobility
	The use of partnerships within the NICE project to intensify the use of information and communication technologies and increase energy efficiency. Working with cities such as Amsterdam, Barcelona, Berlin, Helsinki Manchester, and Rome to enable broadband and fibre networks
	To intensify the use of innovative IT tools in public administration in seven European cities: Helsinki, Berlin, Amsterdam, Paris, Rome, Barcelona, and Bologna. Focus on addressing the low use of IT by urban authorities in nonurban geographically remote or socially excluded areas

Source: own work based on: Rześny-Cieplińska (2018); Rześny-Cieplińska and Wach-Kloskowska (2017) of the urban platform as a form of participation of the urban community and other urban stakeholders in creating the Smart City. This is in line with the idea of the fivefold helix presented in the first chapter as one of the necessary conditions for creating a city in which the improvement of the quality of life includes all inhabitants.

The city of Amsterdam is also engaged in extensive international cooperation for the improvement and development of Smart City solutions, which fosters the popularisation of the Smart City concept and allows for the realisation of the synergistic effects of this cooperation, in the form of sharing knowledge and experience, co-creating smart solutions, jointly raising funds for smart logistics investments, and sharing resources necessary for the creation of logistics innovations in the city.

In light of the above, it can be concluded that the Amsterdam authorities take the implementation of the Smart City concept very seriously, respecting all aspects raised in this regard in the literature and practice. The identified results of their comprehensive activities also partially refute the argument that the Smart City concept is utopian, as they document the existence of real, tangible solutions to improve the quality of life in the city. At the same time, various representatives of the urban community participate in their planning and implementation.

The second international case study of smart logistics solutions is Barcelona. The actions taken by the authorities of this Spanish city in the area of smart are presented in Table 4.3.

In the case of Barcelona, mobility solutions are supported by IT and ICT technologies, which is typical of urban logistics in a Smart City. Additionally, as in Amsterdam, the commitment to ecological forms of transport and the integration of intelligent transport systems draws attention. It is worth mentioning that their infrastructure also includes solutions for people with disabilities, which emphasises the sustainability of the concept implemented by the city authorities. The element of community participation is realised through the *Open Management* platform and universal, free access to the Internet. The data in Table 4.3 also show that Barcelona is engaging in intensive cooperation in Smart City development and is present in many joint projects with Amsterdam and Copenhagen.

Table 4.3 City logistics actions in Barcelona (Spain)

Description of activities	Results of actions
Change of street lighting system from traditional to adaptive LED type	Implementation of solutions in the area of the Internet of Things. Reducing energy consumption in the city Improved efficiency by reducing energy consumption costs
Guidance system to help find parking spaces	Installing electromagnetic sensors in the carriageway, sending information to the app about whether a space is free or occupied Improving the lives of residents and tourists Increased community satisfaction by minimising the time spent searching for available parking spaces
Improving urban freight flows	Night-time delivery systems including noise reduction Control and limit the entry of commercial vehicles in pedestrian priority zones Multifunctional lanes Use of microplatforms for reloading goods onto electric cars or bicycles with cargo space for last-mile delivery
Implementation of projects and applications facilitating transport for residents	Integrated ticket in public transport to facilitate travel by different means—Transports Metropolitan de Barcelona (TMB). Hybrid vehicles in public transport, including those using solar panels Publicly accessible urban bikes as part of the Bicing system Creation of an urban platform within Open Management to inform and enable participation Smart intersections, equipped with devices to help blind people cross the street and minimise the noise created by these devices Implementation of an open city portal for citizens Enabling free internet access for all city residents
Waste management	Installation of smart sensors in waste bins, which has improved waste collection and minimised the inconvenience of waste disposal (overflowing bins; unpleasant odour) In the implemented system, automated waste bins suck the waste into underground storage In addition, the burning of rubbish is used to produce energy used for heating

Source: own work based on Bruska (2012); Antoszewski (2010); Sikora-Fernandez (2016)

In Barcelona, smart urban solutions are also being implemented in waste management, and these are not limited to the installation of sensors that send data and information to the city's cleaning companies, but also include waste treatment that complies with the principles of a closed-loop economy, as the heat obtained from the waste incineration process is used to heat residential and industrial buildings.

In line with this, Barcelona—following the example of other leading European cities—is improving the quality of life of its residents through a number of convenient, smart urban solutions for mobility and waste management.

The third of the analysed cities is Copenhagen, which is famous for the best-developed cycling transport in the world. It is estimated that in this city as much as 80% of the journeys of inhabitants are made by bicycle, on foot, or by public transport. As a result, Copenhagen is significantly reducing its carbon dioxide emissions, in line with the *A Responsible City* project which states that the Danish city will achieve climate neutrality by 2025. Achieving this ambitious target will be unique in the world. Other smart mobility measures in Copenhagen are shown in Table 4.4.

The information in Table 4.4 shows that the municipality even uses bicycles for freight transport. Moreover, in this Smart City, residents are able to benefit from the implementation of modern parking, lighting, and traffic control systems. A unique solution is the Smart Street project, where residents can experience live Smart City technology. This is a very good idea not only to disseminate Smart City solutions but also to educate residents in this field.

It should also be added that Copenhagen's publicity campaign uses three short slogans to express the city's aspirations. The first is the already mentioned *A Responsible City*. The other two are *A City with an Edge* and *A Liveable City*. They refer in turn to:

- maximising the diversity of urban life, including cultural and architectural diversity, which gives the city its uniquely original character,
- improve the quality of life in the city, especially in terms of population growth.

Table 4.4 City logistics action in Copenhagen (Denmark)

Description of activities	Results of actions
Greening the transport	Developing bicycle transport (gradual widening of existing bicycle lanes, reconstruction of all junctions in the city, unification of the colours of the lanes, campaigns promoting correct behaviour on the lanes, intelligent lighting system, switching on to light way for passing cyclists)
Improving urban freight flows	Freight transport by bicycle
Implementation of projects and applications facilitating transport for residents	An integrated transport system is designed to minimise connection time
	Intelligent traffic management system
	Dynamic traffic information system
	Intelligent parking systems (community-based parking)
	Street Lab—an urban space for Smart City solutions to showcase the latest Smart City and Internet of Things technologies
	An open data project called City Data Exchange to foster innovation, stimulate economic activity, and contribute to the success of the CO_2 neutralisation programme

Source: own work based on: Rześny-Cieplińska (2018); Dziura (2017)

The described initiatives and ideas clearly underline the humanistic and sustainable character of Copenhagen's activities as a Smart City that focuses not only on the use of state-of-the-art IT and ICT technologies, but above all on the quality of life of all inhabitants and the greening of urban life.

Comparing and summarising the international perspective on logistics solutions in selected European smart cities, it should be stated that the urban authorities managing them are trying their best to fulfil the recommendations for creating smart, humane, and sustainable places to live. The systems they implement are characterised by a high level of technological sophistication, but are geared towards meeting the needs of the urban community and respecting the environment. This demonstrates a fully conscious and responsible implementation of the SC concept. The observations made, therefore, entitle us to state that the characterised cities represent the highest development generation referred to in the literature and practice as 3.0.

It is worth adding, however, that the implementation of such extensive activities in the field of Smart Citys requires large capital expenditures and the ability to financially maintain the created infrastructure, which for the cities which are less developed economically and less rich in financial resources may be the most serious barrier to the implementation of the most modern technologies. It should be noted and emphasised, however, that the examined cities do not use only their own and internal sources of financing. The financing structure of smart solutions is usually hybrid. For the realisation of the Smart City concept, the analysed entities use not only their own budgetary sources, but also funds from international projects and public-private partnerships, which allow the realisation of very ambitious undertakings, including pro-social and pro-environmental ones, with much less involvement of own financial resources. As a result, commercial operators of such initiatives gain an attractive and permanent form of promotion with a wide international range.

In Poland Intelligent Transport Systems were developed mainly, thanks to financial support from the European Union. It was granted to cities such as Białystok, Bydgoszcz, Trójmiasto, Lublin, Łódź, Kalisz, Koszalin, Kraków, Poznań, Rzeszów, Warsaw, Wrocław, Szczecin, as well as the Transport Association of the Upper Silesian Industrial District (Tomaszewska, 2015). The above-mentioned entities were beneficiaries of the Operational Programme Infrastructure and Environment (OPI&E) Priority VIII: Transport safety and national transport networks Measure 8.3. The following discussion will present case studies relating to Białystok, Rzeszów, and Gdańsk—the cities most frequently and most highly ranked in international Smart City rankings.

Thus, the scope for implementing smart logistics solutions in Białystok is presented in Table 4.5.

According to the information in Table 4.5, the EU support received in Białystok has been allocated to key investments in public transport. This transport has been modernised and environmentally friendly solutions have been implemented. In addition, Białystok has implemented a number of passenger amenities in the form of electronic ticketing and a modern passenger information system. Białystok has also used intelligent transport systems to make freight and passenger flow more flexible.

Table 4.5 City logistics activities in Białystok (Poland)

Description of activities	Results of actions
Start of the project "Improving the functioning quality of the public transport system of the city of Białystok"	Routing on bus lanes, intersections with bus priority arrangements. Introduction of low-floor and low-emission rolling stock equipped with a voice announcement system Implementation of video surveillance and visual passenger information systems
Launching an integrated system introducing the Białystok City Card and dynamic passenger information in electronic communication	Possibility of topping up the card via the internet Adaptation of the transport offer to the needs of the inhabitants Reducing the costs associated with printing and distributing tickets
Improvement of vehicle traffic in cooperation with Simens Sp. z o.o. Introduction of an intelligent transport system covering the whole city of Białystok	Implementation of traffic signal control at 120 intersections. Introducing priority crossing for public transport buses Launching of variable message system boards informing about traffic obstructions, detours, and accidents Creation of a Traffic Management Centre allowing remote operation of all elements As a result, the travel time through the city will be shortened and the city centre will be connected to the other parts of the city and to the neighbouring municipalities belonging to the Białystok agglomeration
Integrated municipal waste management system for Białystok agglomeration financed by the European Union	Construction of a municipal waste thermal treatment plant for the city of Białystok and nine neighbouring municipalities with which the city has entered into municipal partnership agreements. The comprehensive municipal management system includes selective collection of municipal waste, municipal waste processing, including with the use of a modern sorting plant for selectively collected waste (with the application of optical separation), a composting plant for the management of green waste, and an incineration plant for the processing of so-called residual waste The modern waste incineration plant operates in cogeneration mode, i.e., electricity and heat are generated from the incinerated municipal waste with minimum air pollution, thanks to the use of semi-dry flue gas cleaning technology, which combines several functions in a single device: gas absorption of hydrogen chloride, hydrogen fluoride and sulphur dioxide, removal of heavy metals, dioxins, furans and particulates using activated carbon and lime, and flue gas dedusting using a bag filter

Source: own work based on Tomaszewska (2015); Wąsowicz et al. (2018); Ogrodnik and Kolendo (2021)

Nevertheless, it should be noted that the scope of implemented changes and improvements in Białystok is much smaller than in the previously described international case studies, which is due to both the level of development of the Polish economy and the limited resources for urban investment.

It is also worth emphasising that the authorities of Białystok are active in the area of waste logistics. The project has made it possible not only to improve municipal management, but also to significantly green it by reducing the harmfulness of waste to the environment and increasing the level of its recycling. In addition, the modern waste incineration plant was designed in accordance with the highest international standards and respect for the principles of a closed-circuit economy. This means that Białystok is well on the way to achieving the next stages of Smart City development.

A significant range of intelligent logistics solutions have been implemented and used in Rzeszów—Table 4.6. The city has, like Białystok, modernised its transport fleet, introduced logistics facilities for public transport passengers, and installed an intelligent transport system that controls traffic and makes travel time and modes more flexible. An additional logistics solution in this case is the weighing system for freight vehicles, which facilitates the identification of exceeding permissible standards and prevents overexploitation of city roads.

In Gdańsk, the last of the examined cities, smart logistics solutions very similar to those of the two previously described Polish cities were implemented and used. Their overview is presented in Table 4.7.

The city has focused on maximising the use of intelligent logistics systems in both freight and passenger transport. An important priority for Gdańsk is also the greening transport, as evidenced by the expansion of cycle paths and the promotion of electric transport.

From the presented case studies of Polish smart cities, a lower degree of sophistication of smart logistics solutions clearly emerges, which are limited to basic transport systems using IT and ICT technologies (Masik et al., 2021; Dohn et al., 2020; Sikora-Fernandez, 2018). However, it is worth noting that according to the Smart City concept, these systems are integrated with each other and serve to improve the quality of life of residents (Knop & Kramarz, 2020). There are ecological themes in transport

Table 4.6 City logistics activities in Rzeszów (Poland)

Description of activities	Results of actions
Introduction of dynamic passenger Information system	Installation of information boards at bus stops displaying current information on a given course
	As a result, a passenger is served and informed more efficiently
Implementation of a public transport management system	Enabling the location of vehicles to be identified, counting the number of passengers, and controlling the quality standards of passenger service by operators
	This has made it possible to respond to the current needs of public transport in terms of the timetable reconstruction of the road system, verification of control algorithms
	Extending the urban road network, making wider pavements, creating cycle paths, and setting aside bus lanes to facilitate the movement of all residents
	Purchase of modern buses that meet the EEV emission standard (30 12-m diesel buses, 30 12-m natural gas buses, and 20 10-m diesel buses)
Use of electronic ticket for public transport	Introduction of e-ticketing
	Use of electronic form of payment for bus travel on Rzeszów City Card
	Creating the possibility to manage the e-ticket with the help of stationary and bus-mounted ticketing machines, ticket punchers, or via a web service regardless of the day and time
Use of area traffic control	Implementation of a comprehensive solution consisting of (1) a traffic light control system to maintain traffic flow and minimise waiting times at intersections; (2) a public transport vehicle priority system, assigning priority to public transport; and (3) a driver information system using variable message system signs (obstructions, changes in traffic organisation, recommended detours)
	These systems are connected by radio and thus provide communication with traffic management centres, with public transport, and with bus stops and city buses
Implementation of the dynamic vehicle weighing system	Identification of traffic of vehicles exceeding weight standards by means of measuring points equipped with weighing plates and variable message system signs

Source: own work based on: Nowotyńska and Kut (2016); Tompalski and Leśniak (2018)

Table 4.7 City logistics activities in Gdańsk (Poland)

Description of activities	Results of actions
Greening the transport	Developed system of bicycle paths
	Promoting electric and alternative energy vehicles
	Implementation of a street lighting management system
Improving urban freight flows	Temporary and tonnage restrictions on trucks entering the city
	Introduction of road bays for loading and unloading
Implementation of projects and applications facilitating transport for residents	Implementation of the TRISTAR system as an intelligent traffic management system. This system includes two main modules:
	• public transport management system (information for public transport passengers; traffic management for road transport vehicles)
	• urban traffic management system (traffic control; traffic monitoring; video surveillance; meteorological measurement; driver information; parking information; road safety management)
	Use of the ACCUS platform for coordination and integration of urban systems
	Integrated charging system for entering certain parts of the city.
	Introduction of an automatic tolling system when entering certain zones of the city

Source: own work based on: Rześny-Cieplińska (2018); Rześny-Cieplińska and Wach-Kloskowska (2017)

investments in the form of the use of low-emission vehicles in urban transport and the creation of cycle paths; nevertheless, they are definitely more modest in scope than in the case of the described European cities. There is also a lack of public participation in the solutions used. On the basis of those observations, it is possible to confirm the conclusion indicated in the first chapter that Polish entities aspiring to become smart represent generation 1.0 cities and should be much more active in the ecological and social areas of Smart City development.

Unfortunately, due to the relatively poor development and the state of the road infrastructure, there are many ongoing challenges in Poland which consume most of the efforts and resources of city authorities, making the transition to a higher stage of SC development very difficult and will continue to be so. This is also evidenced by the fact that the cases described, as well as other Polish smart cities, have managed to implement smart logistics solutions mainly, thanks to 80% project-based funding from the European Union. Without this support, the implementation of the described logistics measures would probably not have been possible at all.

In Poland, investments in logistics infrastructure are financed almost entirely from public funds using European Union subsidies. A significant number of cities and villages face a budget deficit and significant debt, so the main barrier to the development of smart cities is economic and financial in nature. Another obstacle is the low level of social participation and environmental awareness, which significantly complicates the development of the social and environmental aspects of a Smart City by city authorities. Examples of European cities, however, clearly prove that change for the better is possible and the concept of smart cities can be successfully put into practice.

References

Antoszewski, M. (2010). Barcelona—europejska stolica innowacji. *Dell Technologies. Rozwiązania i usługi.* https://www.delltechnologies.com/pl-pl/blog/barcelona-europejska-stolica-innowacji/. Accessed 21 July 2021

Aparicio, A. (2020). Streamlining the implementation process of urban mobility innovations: Lessons from the ECCENTRIC project in Madrid. *Transport Policy, 98*, 160–169.

Ashwin, M., Alqahtani, A., & Mubarakali, A. (2021). Iot based intelligent route selection of wastage segregation for smart cities using solar energy. *Sustainable Energy Technologies and Assessments, 46*, 1–9.

Bellandi, M., Monti, A., Scerbo, M., Colomer, J. V., Colomer, O., & Fiore, M. (2014). Polis-the experience of the Tuscan innovation cluster in the field of sustainable mobility. *Procedia – Social and Behavioral Sciences,* *162,* 398–407.

Bruska, A. (2012). Logistyka jako komponent Smart City. *Studia Miejskie, 6,* 9–19.

Calantone, R., Cavusgil, S., & Zhao, Y. (2002). Learning orientation, firm innovation capability, and firm performance. *Industrial Marketing Management, 31*(6), 515–524.

Chen, C., & Pan, S. (2015). *Using the crowd of taxis to last mile delivery in e-commerce: a methodological research. Conference: SOHOMA15* (pp. 61–70). University of Cambridge.

Cichosz, M., Nowicka, K., & Pluta-Zaremba, A. (2014). Innowacje w zarządzaniu transportem w miastach. In M. Bryx (Ed.), *Innowacje w zarządzaniu miastami w Polsce.* Oficyna Wydawnicza SGH.

Digiesi, S., Fanti, M. P., Mummolo, G., & Silvestri, B. (2017). Externalities reduction strategies in last mile logistics: a review. *International conference on service operations and logistics, and informatics, Bari (Italy)* (pp. 248–253).

Dohn, K., Kramarz, M., & Przybylska, E. (2020). Luki i aspekty logistyczne w strategiach polskich miast. In I. Jonek-Kowalska & J. Kaźmierczak (Eds.), *Inteligentny rozwój inteligentnych miast* (pp. 55–72). CeDeWu.

Dziura, Ł. (2017). Dobre praktyki w zakresie smart city. In J. Czapska, P. Mączyński, K. Struzińska, & J. A. K. Kraków (Eds.), *Bezpieczne miasto. W poszukiwaniu wiedzy przydatnej praktykom* (pp. 148–168).

Ferrer, J., & Alba, E. (2019). BIN-CT: Urban waste collection based on predicting the container fill level. *BioSystems, 186,* 1–9.

Fijałkowski, J. (2010). Centrum konsolidacji ładunków dla obsługi logistycznej miasta. Prace naukowe Politechniki Warszawskiej. *Transport, 76,* 33–42.

Grad, B., Ferensztajn-Galardos, E., & Krajewska, R. (2013). Innowacyjne rozwiązania w miejskim transporcie zbiorowym na przykładzie Radomia. *Transport Miejski i Regionalny, 3,* 13–18.

Heitz, A., & Beziat, A. (2016). The parcel industry in the spatial organization of logistics activities in the Paris Region: inherited spatial patterns and innovations in urban logistics systems. *Transportation Research Procedia, 12,* 812–824.

Herbuś, I. (2015). Innowacje w miastach jako wyznacznik sukcesu współczesnych samorządów. Zeszyty Naukowe Politechniki Częstochowskiej. *Zarządzanie, 19,* 35–43.

Hirschhorna, F., Paulssonb, A., Sørensen, C. H., & Veeneman, W. (2019). Public transport regimes and mobility as a service: Governance approaches in

Amsterdam, Birmingham, and Helsinki. *Transportation Research Part A, 130*, 178–191.

Iwan, S., Allesch, J., Celebi, D., Kijewska, K., Hoé, M., Klauenberg, J., & Zajicek, J. (2018). Electric mobility in European urban freight and logistics—status and attempts of improvement. *Transportation Research Procedia, 39*, 112–123.

Jąderko, K., Stępień, M., Białecka, B. (2016). Wyzwania w projektowaniu innowacyjnych systemów IT w gospodarce odpadami komunalnymi. In: Systemy wspomagania w Inżynierii Produkcji. Metody i narzędzia Inżynierii Produkcji dla rozwoju inteligentnych specjalizacji. (Eds.) Gembalska-Kwiecień, A., 4(16), 43–56.

Jaroszyński, J. W., & Chłąd, M. (2015). Koncepcje logistyki miejskiej w aspekcie zrównoważonego rozwoju. Studia Ekonomiczne. *Zeszyty Naukowe Uniwersytetu Ekonomicznego w Katowicach, 249*, 164–171.

Jin, P., Mangla, S., & Song, M. (2021). Moving towards a sustainable and innovative city: Internal urban traffic accessibility and high-level innovation based on platform monitoring data. *International Journal of Production Economics, 235*, 1–13.

Karnan, H., Ananva, R. S., Sri Sushmithaa, V., & Vinothine, G. (2021). Smart arm for segregation of biomedical waste. *Materials Today: Proceedings*, 1–7.

Karvonen, A., Evans, J., & van Heur, B. (2014). The politics of urban experiments: Radical change or business as usual? In S. Marvin & M. Hodson (Eds.), *After sustainable cities?* (pp. 105–114). Routledge.

Kauf, S. (2016). Współczesne wyzwania dla logistyki miasta—kształtowanie nowych struktur przestrzennych w dostawach towarów. Zeszyty Naukowe Politechniki Częstochowskiej. Zarządzanie, 24(1), 128–139.

Kiba-Janiak, M. (2012). Wybrane rozwiązania w logistyce miejskiej na rzecz poprawy jakości życia mieszkańców. *Studia Miejskie, 6*, 41–50.

Knop, L., & Kramarz, M. (2020). Attractiveness of the region in connection with intermodal transport development. In K. Grzybowska, A. Awasthi, & R. Sawhney (Eds.), *Sustainable logistics and production in industry 4.0. New opportunities and challenges* (pp. 197–217). Springer.

Kozerska, M., & Konopka, M. (2018). Zastosowanie inteligentnych systemów transportowych w sytuacjach ograniczonego dostępu do miast. *Zeszyty Naukowe Politechniki Śląskiej, seria: Organizacja i Zarządzanie, 130*, 353–363.

Koźlak, A. (2008). Inteligentne systemy transportowe jako instrument poprawy efektywności transportu. *Logistyka, 2.*

Kręt, P. (2020). Inteligentne systemy transportowe w Smart City. *Management and Quality—Zarządzanie i Jakość, 2*(2), 42–54.

Lewicki, A. (2018). Technologie inteligentnego zarządzania gospodarką odpadami wykorzystujące architekturę Internetu rzeczy. In A. Białowiec (Ed.), *Innowacje w gospodarce odpadami* (pp. 141–155). Wydawnictwo Uniwersytetu Przyrodniczego we Wrocławiu.

Mamun, A., Hannan, M. A., Hussain, A., & Basri, H. (2016). Theoretical model and implementation of a real time inteligent bin status monitoring system using rule based decision algorithms. *Expert Systems with Applications, 48*, 76–88.

Marques, P., Manfroi, D., Deitos, E., Cegoni, J., Castilhos, R., Rochol, J., Pignaton, E., & Kunst, R. (2019). An IoT-based smart cities infrastructure architecture applied to a waste management scenario. *Ad Hoc Networks, 87*, 200–208.

Masik, G., Sagan, I., & Scott, J. W. (2021). Smart City strategies and new urban development policies in the Polish context. *Cities, 108*, 102970.

Montwiłł, A., Pietrzak, O., & Pietrzak, K. (2021). The role of Integrated Logistics Centers (ILCs) in modelling the flows of goods in urban areas based on the example of Italy. *Sustainable Cities and Society, 69*, 1–15.

Moreno, R., Paci, R., & Usai, S. (2005). Spatial spillovers and innovation activity in European regions. *Environment & Planning A, 37*(10), 1793–1812.

Müller, S., Rudolph, C., & Janke, C. (2019). Drones for last mile logistics: Baloney or part of the solution? *Transportation Research Procedia, 41*, 73–87.

Nakandhrakumar, R. S., Rameshkumar, P., & Rao, T. (2021). Internet of things (IoT) based system development for robotic waste segregation management. *Materials Today: Proceedings*, 1–5.

Nam, T., & Pardo, T. A. (2011). Smart City as urban innovation: Focusing on management, policy, and context. Paper presented at the ICEGOV 2011, Tallinn, Estonia.

Nathnail, E., Gogas, M., & Adamos, G. (2016). Urban freight terminals: A sustainability cross-case: Analysis. *Transportation Research Procedia*, 394–402.

Niedzielski, P., & Tundys, B. (2012). Benchmarking jako kreator innowacyjności w logistyce miejskiej. *Logistyka, 3*, 1657–1669.

Nilssen, M. (2019). To the smart city and beyond? Developing a typology of smart urban innovation. *Technological Forecasting and Social Change, 142*, 98–104.

Nowicka, K. (2015). Innowacje w logistyce miejskiej—ITS jako usługa. *Research Papers of Wrocław University of Economics, 383*, 108–120.

Nowotyńska, I., & Kut, S. (2016). Nowoczesne systemy transportowe w komunikacji miejskiej. *Logistyka, 12,* 1643–1646.

Ogrodnik, K., & Kolendo, Ł. (2021). Application of gis technology and AHP to determine the areas with fully developed, compact functional and spatial structure: A case study of Białystok, Poland. *Land Use Policy, 109,* 105616.

Pacheco, F., Furtado, F., & Filho, E. (2018). Stepbox: A proposal of share economy transport service. *13th Iberian conference on information systems and technologies (CISTI), Caceres, Spain.*

Pan, S., Chen, C., & Zhong, R. Y. (2015). *A crowdsourcing solution to collect e-commerce reverse flows in metropolitan areas, INCOM 2015, Ottawa* (pp. 1–7).

Patier, D., & Browne, M. (2010). A methodology for the evaluation of urban logistics innovations. *Procedia Social and Behavioral Sciences, 2,* 6229–6241.

Pluta-Zaremba, A. (2015). Innowacje w logistyce miejskiej—zrównoważony transport towarów. *Research papers of Wrocław University of Economics, 383,* 154–165.

Przybylska, E., Kruczek, M., & Żebrucki, Z. (2016). Innowacyjność branży TSL według klasyfikacji OECD. In J. Witkowski & S. Saniuk (Eds.), *Przedsiębiorczość i Zarządzanie. Systemy logistyczne w gospodarowaniu* (pp. 421–432). Wydawnictwo Społecznej Akademii Nauk, 12(II).

Rekha, M., Kodhai, E., Kuzhaloli, S., Sharma, P., Kumar, A., & Kumar, N. (2021). IoT based garbage bin monitoring and decluttering system. *Materials Today: Proceedings,* 1–4.

Rodríguez-Pose, A., & Crescenzi, R. (2008). Research and development, spillovers, innovation systems, and the genesis of regional growth in Europe. *Regional Studies, 42*(1), 51–67.

Rodríguez-Pose, A., Wilkie, C., & Zhang, M. (2021). Innovating in "lagging" cities: A comparative exploration of the dynamics of innovation in Chinese cities. *Applied Geography, 132,* 1–14.

Russo, S. M., Voegl, J., & Hirsch, P. (2021). A multi-method approach to design urban logistics hubs for cooperative use. *Sustainable Cities and Society, 69,* 1–17.

Rześny-Cieplińska, J. (2018). Strategie logistyki miejskiej wobec koncepcji Smart City na przykładzie miast polskich i zachodnioeuropejskich. *Prace Naukowe Uniwersytetu Ekonomicznego we Wrocławiu, 505,* 473–479.

Rześny-Cieplińska, J., & Wach-Kloskowska, M. (2017). Logistyczne aspekty koncepcji Smart City. Studium przypadku na podstawie miast europejskich. *Studia Miejskie, 27,* 129–141.

Serafini, S., Nigro, M., Gatta, V., & Marcucci, E. (2018). Sustainable crowd-shipping using public transport: a case study evaluation in Rome. *Transportation Research Procedia, 30*, 101–110.

Selwon, A., & Roman, K. (2017). Wpływ Inteligentnych Systemów Transportowych na redukcję kongestii w miastach. *Autobusy, 3*, 28–32. https://doi.org/1024136/atest.2017.010

Sikora-Fernandez, D. (2016). Praktyczne aspekty budowy Smart City na przykładzie Barcelony. *Prace Naukowe Uniwersytetu Ekonomicznego we Wrocławiu, 432*, 155–164.

Sikora-Fernandez, D. (2018). Smarter cities in post-socialist country: Example of Poland. *Cities, 78*, 52–59.

Sinha, A., Gupta, K., Jamshed, A., & Singh, R. (2021). Intelligent dustbin: A strategic plan for smart cities. *Materials Today: Proceedings*, 1–6.

Sonn, J. W., & Storper, M. (2008). The increasing importance of geographical proximity in knowledge production: An analysis of US patent citations, 1975-1997. *Environment & Planning A, 40*, 1020–1039.

Standing, C., Biermann, S., & Standing, S. (2018). The implications of the sharing economy for transport. *Transport Reviews*, 1–17.

Szelągowska, A. (2014). Definicyjne aspekty innowacji. In M. Bryx (Ed.), *Innowacje w zarządzaniu miastami w Polsce*. Oficyna Wydawnicza SGH.

Szołtysek, J. (2016). Ekonomia współdzielenia a logistyka miasta—rozważania o związkach. *Gospodarka Materiałowa i Logistyka, 11*, 2–9.

Szromnik, A. (2012). Miasto przedsiębiorcze i innowacyjne w regionie. Zeszyty Naukowe Uniwersytetu Szczecińskiego. *Ekonomiczne problemy usług, 98*, 323–346.

Taniguchi, E., Thompson, R. G., & Qureshi, A. G. (2020). Modelling city logistics using recent innovative technologies. *Transportation Research Procedia, 46*, 3–12.

Taniguchi, E., Thompson, R. G., & Yamada, T. (2014). Recent trends and innovations in modelling city logistics. *Procedia – Social and Behavioral Sciences, 125*, 4–14.

Tomaszewska, E. J. (2015). Zeszyty Naukowe Uniwersytetu Szczecińskiego, 875: Problemy Zarządzania, Finansów i Marketingu, 41, 2, 318–329. https://doi.org/10.18276/pzfm.2015.41/2-26.

Tompalski, K., & Leśniak, A. (2018). Inteligentne systemy transportowe i ich zastosowanie w mieście Rzeszowie. *Zeszyty Studenckie Wydziału Ekonomicznego "Nasze Studia", 9*, 68–77.

Van Duin, J. H. R., Quak, H. J., Anand, N., & van den Band, N. (2020). Designing sharing logistics as a disruptive innovation in city logistics. *4th international conference green cities 2020—green logistics for Greener Cities, 3–5 June 2020, Szczecin, Poland* (pp. 1–11).

Verlinde, S., & Macharis, C. (2016). Innovation in urban freight transport: the triple helix model. *Transportation Research Procedia, 14*, 1250–1259.

Viu-Roig, M., & Alvarez-Palau, E. J. (2020). The impact of e-commerce-related last-mile logistics on cities: A systematic literature review. *Sustainability, 12*, 1–19.

Vural, C., & Aktepe, C. (2021). Why do some sustainable urban logistics innovations fail? The case of collection and delivery points. *Research in Transportation Business & Management*, 1–12.

Wąsowicz, K., Famielec, S., & Chełkowski, M. (2018). *Gospodarka odpadami komunalnymi we współczesnych miastach*. Kraków.

Wojewódzka-Król, K., & Rolbiecki, R., (2010). Inteligentne systemy transportowe w świetle europejskiej polityki transportowej. In *E-gospodarka w Polsce, Stan obecny i perspektywy rozwoju, cz. I, Zeszyty Naukowe Uniwersytetu Szczecińskiego nr 597, Ekonomiczne Problemy Usług nr 57, Wyd* (p. 70). Naukowe Uniwersytetu Szczecińskiego, Szczecin.

Ydersbonda, I. M., Auvinenb, H., Tuominenb, A., Fearnleya, N., & Aarhauga, J. (2020). Nordic experiences with smart mobility: Emerging services and regulatory frameworks. *Transportation Research Procedia, 49*, 130–144.

5

Methodology for Assessing the Impact of City Logistics Maturity on the Level of City Intelligence

Methodology for Assessing the Advancement of Cities in Terms of Smart City Solutions

The assessment of the city in the context of being "smart" from the methodological point of view was carried out primarily on the basis of the considerations presented in the theoretical part of the monograph relating to the Smart City concept. It also uses practical observations obtained during analysing case studies of Polish and foreign smart cities.

And so, the universal classification of smart cities uses two key literature threads relating to the level of advancement of the city in the implementation of the Smart City concept (Taratori et al., 2021; Schütz et al., 2019; Fernandez-Anez et al., 2018; McAdam & Debackere, 2018; Calzada, 2017; Girardi & Temporelli, 2017; Lee et al., 2014):

- The thread on the successive economic helices: triple (business-city-science), quadruple (business-city-science-local community), and five-fold (business-city-science-local community-ecological organisations),
- The thread concerns the next stages of Smart City development, including the generation of cities defined as 1.0, 2.0, and 3.0.

On this basis, seven levels were distinguished synthetically in Table 5.1.

The categorisation presented in Table 5.1 at levels 4–6 directly relates to the evolution of smart cities and their next generations and can be applied to cities that have already achieved the smart status. On the other hand, levels from 0 to 3 allow to determine the advancement of cities on the path leading to being smart. It is important because many—not only Polish cities—are struggling with serious economic, social, civilisation, and technological problems and despite their interest in the development of Smart City solutions, these entities are not able to fully engage in their implementation. The review of the case studies presented in this chapter and the research conducted so far (Jonek-Kowalska and Wolniak, 2021; Wolniak and Jonek-Kowalska, 2021) clearly show significant differences in the level of advancement of intelligent solutions in Polish cities and in reference European cities. Therefore, limiting the assessment of the level of advancement of intelligent solutions in cities to their next generations 1.0, 2.0, and 3.0 would flatten the assessments of cities that do not fit into the characteristics of the above generations, and make efforts to be smart. This allows for the diversification and assessment of the advancement level of those cities that are just looking for their path in the implementation of the Smart City concept, which is of particular importance in relation to Polish cities being the subject of research in this monograph. As it has already been emphasised, those Polish cities that actively and consciously implement the Smart City concept can be classified at most as generation 1.0 cities, and most of them have not even reached this level yet.

Taking into account the presented observations and the results of the research conducted so far, in the questionnaire in part concerning the assessment of the advancement of Polish cities in the implementation of the Smart City concept, six questions were formulated relating to the distinguishing features of individual levels described in Table 5.1. At the same time, taking into account the fact that there are no 3.0 generation smart cities in Poland, the study did not include cooperation with environmental organisations, so as not to create an empty analytical category. The list of questions with the corresponding discriminants is presented in Table 5.2.

Table 5.1 Level of advancement of intelligent solutions in cities

Level	Characteristics
Zero (0)	**The city does not know the Smart City concept and is not interested in its implementation**
	It is a level that bodes badly for the possibility of implementing Smart City solutions, proving that the city authorities ignore development goals and are ignorant of the latest strategic trends. Even if there are serious economic or social problems in a city with such characteristics, it does not justify the lack of knowledge, and thus the lack of development vision, even of the most distant one. For these reasons, this level has been assigned the number "0" and the name zero
The first (1)	**The city knows the Smart City concept, but is not interested in implementing it**
	A city representing this level focuses on current activities, which usually result from serious economic problems and the accompanying low quality of life of its inhabitants. Although it is familiar with the concept of a Smart City, it does not have the time and financial resources to engage in its implementation. It treats it as a very distant development vision, which ultimately does not preclude its implementation in the long term
The second (2)	**The city knows the concept of Smart City, is interested in its implementation, and includes it or plans to include it as part of the development strategy**
	This is the level at which the conscious creation of Smart City solutions begins. The city authorities are aware of their existence and want to provide them to inhabitants to improve their quality of life
The third (3)	**The city knows the Smart City concept, is interested in its implementation, and includes it in its development strategy, and evaluates the level of being "smart" in accordance with international standards in this area**
	At this level, the city not only implements the Smart City solutions included in the strategy, but also assesses itself and compares to other cities considered smart. Such a diagnosis allows to set the desired development directions and focuses the implementation of the Smart City concept on specific actions

(continued)

Table 5.1 (continued)

Level	Characteristics
The fourth (4)	**The city knows the Smart City concept, is interested in its implementation, and includes it in its development strategy, and evaluates the level of being "smart" in accordance with international standards in this area. Additionally, it is interested in cooperation with representatives of the science sector to implement Smart City solutions**
	On this level, an aware and active creation of Smart City solutions begins. A city that meets the above conditions operates according to a triple helix involving the cooperation of business, city authorities, and representatives of the science sector. On the path of evolution of smart cities, it represents generation 1.0
The fifth (5)	**The city knows the Smart City concept, is interested in its implementation, and includes it in its development strategy, and evaluates the level of being "smart" in accordance with international standards in this area. Additionally, it is interested in cooperation with representatives of the science sector to implement Smart City solutions and identify the local society needs**
	This level is a development of the scope of cooperation on the line: business-city science with the local community and is characteristic of the economic quadruple helix. On the path of evolution of smart cities, such entity represents generation 2.0
The sixth (6)	**The city knows the Smart City concept, is interested in its implementation, and includes it in its development strategy, and evaluates the level of being "smart" in accordance with international standards in this area. Additionally, it is interested in cooperation with representatives of the science sector to implement Smart City solutions and identify the local society needs. It also takes steps to engage environmental organisations in building smart cities**
	At this stage, the city collaborates with all stakeholders in the creation of intelligent solutions according to the fivefold helix (business-city-science-local community-ecological organisations). The benefits of this cooperation are maximum and multifaceted. On the path of evolution of smart cities, such entity represents generation 3.0

Source: own work

Table 5.2 Survey questions diagnosing the level of advancement of intelligent solutions in Polish cities

Feature	Question
Knowledge of the SC concept	Are the city authorities **familiar with the concept** of a Smart City?
Implementation of the SC concept	Would your city be interested in **obtaining the status of a Smart City** in the currently emerging and widely disseminated international rankings?
Including the concept of SC in the strategy	Are you planning to **include the Smart City concept in the city's strategy**?
Evaluation (self-assessment) in terms of being smart	Does your city make a **comprehensive self-assessment** of "being" a Smart City?
Cooperation with the science sector	Do you assume the **possibility of cooperation** in the field of Smart City planning and development with **research and academic units** specialising in the implementation of Smart City solutions?
Identification of the needs and expectations of local communities	Do you see the need to **identify the needs and expectations of local communities** in the field of smart solutions?

Source: own work

The cafeteria of answers to the above questions included three answers: "Yes", "no", and "I have no opinion". Such a system of answers creates several possible configurations describing the advancement level of Polish cities in the implementation of the Smart City concept. The rules for assigning them to the previously distinguished levels (from 0 to 5) are presented in Fig. 5.1. And yes, the transition to the fifth level requires that you answer "yes" to all six questions. For the first level, it is necessary to answer the first two questions. The second level requires an affirmative answer to the first three questions, etc. If the answer to one of the following questions is "no" or "I have no opinion", then the level is determined by the last of the questions to which the answer was "yes" (the city does not go to the next level regardless of the layout of the answers to further questions). The presented approach allows to divide the studied cities into groups according to the level of advancement in the implementation of the Smart City concept. This, in turn, will be the starting point for analysing the relationship between the degree of advancement of smart solutions in cities and the logistics maturity of the cities studied.

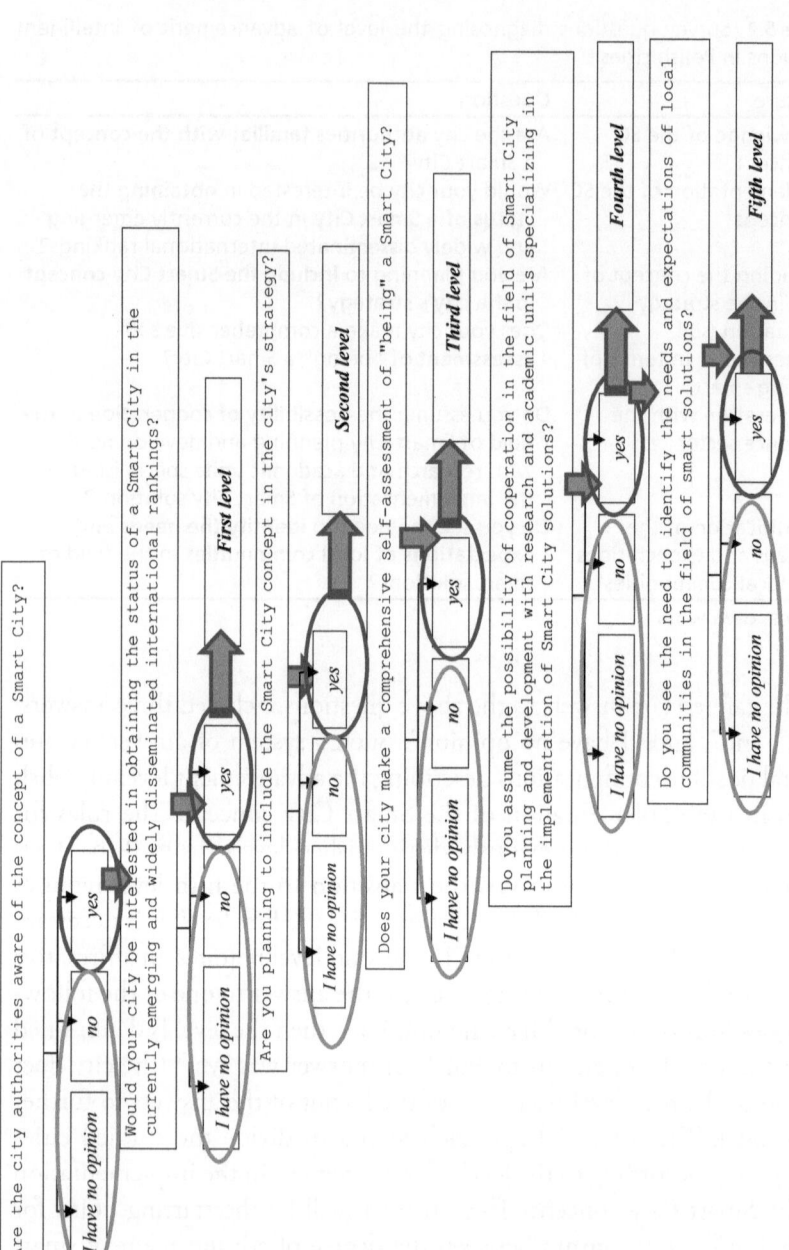

Fig. 5.1 Principles of assessing the level of advancement of intelligent solutions in Polish cities in the context of conducted surveys. Source: own work Assumptions and the concept of the assessment of logistics maturity of Polish cities

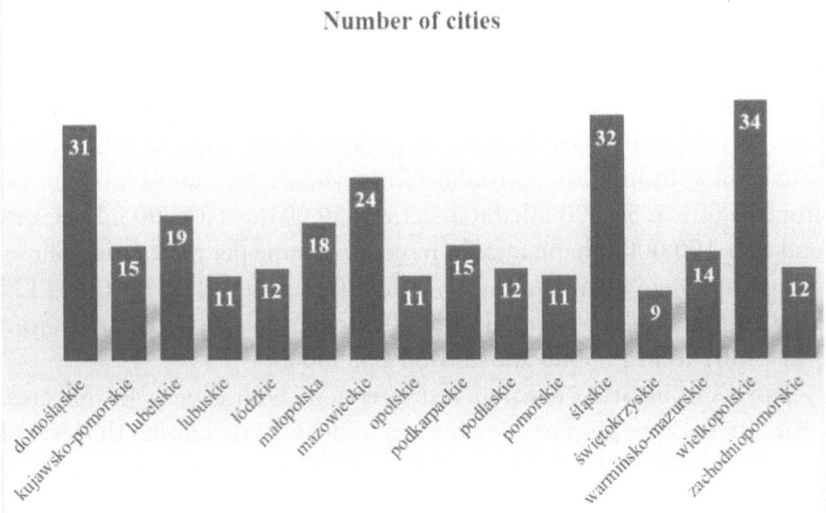

Fig. 5.2 Structure of the examined cities by voivodships. Source: own work

Research on logistics maturity was carried out on a sample of 280 Polish cities in the second half of 2019, similar to the research conducted on the level of advancement of intelligent solutions in Polish cities. The selection of the sample for research was random, assuming a confidence level of 0.95, a maximum error of 5%, and a fraction of 0.5. On the other hand, the selection of cities in individual voivodeships was layered. The structure of the examined cities is shown in Fig. 5.2.

The research was carried out using the Computer-Assisted Telephone Interviewing (CATI) method and, in some cases, the Computer-Assisted Web Interview (CAWI) method by completing an electronic questionnaire as part of a computer-assisted interview with the use of a website. The questionnaire was addressed to people responsible for developing strategies and plans for the development of cities. Comprehensive research covered various issues regarding the assessment of the potential for the development of smart cities in Poland, while the results presented in this publication concern some of the analysed issues.

The key to achieving the objectives of the monograph is the studies that concerned the determination of the degree of logistics maturity of the cities studied along with the assessment of its dependence on the basic measures of statistical description for the entire population under study, as well as the division of the cities studied according to their size (number of inhabitants) and income per capita. Regarding the size of the city, the following have been distinguished: units with less than 5000 inhabitants, from 5001 to 10,000, from 10,001 to 25,000 inhabitants, from 25,001 to 50,000 inhabitants, from 50,001 to 100,000 inhabitants and over 100,000 inhabitants. In terms of income per capita, the following groups were identified: below PLN 1000, from PLN 1000 to PLN 2000, from PLN 2001 to PLN 3000, from PLN 3001 to 4000, from PLN 4001 to PLN 5000 and above PLN 5000.

Based on literature research and interviews with experts in this area, four criteria were proposed for the respondents to determine the level of logistics maturity of the city:

- the level of recognition of logistic areas in the formulated strategies of the cities studied,
- the level of cooperation between the managers of the examined cities and the stakeholders of city logistics,
- the scope and timing of activities in the field of city logistics,
- the level of implementation of intelligent solutions in the field of city logistics.

Each of the criteria represents a broad approach to the examined issue. This allows for a full and comprehensive approach to research problem—determining levels of logistics maturity of cities. Details of the criteria, by indicating the aspects examined within them, are presented in Table 5.3.

As part of the research, each of the areas presented in Table 5.3 was assessed separately by the managers of all 280 cities. A five-point scale from 0 to 4 points was adopted. As part of each of the criteria, the points obtained for all areas assigned to it within a given city were summed up. Then, the obtained score was presented as a percentage. The final assessment of the criterion, which is the basis for the subsequent assignment of the level of logistics maturity of the city, is presented in formula (5.1):

Table 5.3 Criteria for determining the level of logistics maturity of cities

No.	Criterion name	Aspects/areas covered by the criterion
1	The level of recognition of logistic areas in the formulated strategies of the cities studied	Collective passenger transport
		Individual car transport
		Freight transport
		Location of logistics/distribution centres
		Location of trading and production enterprises and the way of their communication
		Location of recreational areas and the way they are connected
		Location of cultural organisations and the way in which they are connected
		Location of health-related organisations and how they are connected
		Location of inhabitantial zones and the way of their communication
		Location of organisations related to the safety of inhabitants and the manner of their communication
		Broadly understood cooperation in the field of creating innovative logistics solutions in the city
		Waste management
		Circular economy
		Low mobility
		Movement of elderly and disabled people
2	The level of cooperation between the managers of the examined cities and the stakeholders of city logistics	Inhabitants
		Production companies
		Trading companies
		Transport and logistics companies
		Organisations related to public safety
		Organisations related to waste management

(continued)

Table 5.3 (continued)

No.	Criterion name	Aspects/areas covered by the criterion
3	The scope and timing of activities in the field of city logistics	Waste segregation in the city
		Collection of oversized electro-waste clothes, etc.
		Free travel by public transport
		Low mobility
		Car-pooling
		Car-sharing
		The use of publicly available information systems in the field of passenger transport
		The use of publicly available information systems in the field of freight transport
		The use of publicly available information systems in the field of transport of elderly and disabled persons
		Promoting ecological means of transport
4	The level of implementation of intelligent solutions in the field of city logistics	Intelligent waste management systems
		Solutions limiting freight road transport
		Solutions to reduce CO_2 emission
		Noise reduction solutions
		Intelligent solutions for excluded people
		Intelligent passenger information systems
		Intelligent traffic control systems
		Low-mobility solutions
		Intelligent street lighting organisation
		Information about free parking spaces in the city in real time
		Technologies of accessibility and timeliness of public information
		Air quality monitoring technologies
		Modern geographic information systems
		Internet of things

Source: own work

$$Ok = \frac{\sum_{i=1}^{n} Oo}{Ok_{max}} \times 100\% \qquad (5.1)$$

where:

Ok—total evaluation of a given criterion,

n—number of areas examined under a given criterion,

Oo—partial assessment was obtained for each area within a given criterion,

Ok_{max}—the maximum possible score for the entire criterion (5.2), where:

$$Ok_{max} = 4 \times n \qquad (5.2)$$

where

4—Maximum Score for each Criterion

n—number of areas examined under a given criterion.

The assessments of criteria determined in this way, separately for each city under study, were compared to the assumed levels of logistics maturity. There were five levels of maturity for which the percentage ranges necessary to achieve a given level were proportionately defined. The scoring scale used for individual criteria as well as the adopted ranges for the five levels of logistics maturity is presented in Table 5.4.

Based on the formulated assumptions, each of the 280 cities was initially assigned to one of the five maturity levels, separately for each criterion. In the next phase of the research, the final level of the city's logistics maturity was determined. For this purpose, it was assumed that the city only moves to a higher level of logistics maturity when it reaches this level under each of the four criteria. Thus, it can be assumed that the final level of logistics maturity of the city, being the resultant of all four criteria, is the maximum level achieved by all examined criteria.

Table 5.4 Assumptions for determining the level of logistics maturity of cities

No.	Criterion	The adopted grading scale [points]	Assessment intervals for individual levels of logistics maturity [%]				
			1	2	3	4	5
1	The level of recognition of logistic areas in the formulated strategies of the cities studied	0—The area is not included 1—The area is not included 2—The area is minimally covered 3—The area is covered on a general level 4—The area is covered in detail	<0–20)	<20–40)	<40–60)	<60–80)	<80–100>
2	The level of cooperation between the managers of the examined cities and the stakeholders of city logistics	0—No interaction with the stakeholder 1—Conducting research on the needs of a given stakeholder 2—Trying to involve the stakeholder in the search for new solutions 3—Indicating to the stakeholder their needs and potential opportunities to meet them 4—Permanent stakeholder inclusion in the strategic decision-making process	<0–20)	<20–40)	<40–60)	<60–80)	<80–100>
3	The scope and timing of activities in the field of city logistics	0—The action was never taken 1—The action was undertaken in the last year 2—The action was undertaken between 1 and 3 years ago 3—The action was undertaken between 3 and 5 years ago 4—The action was taken more than 5 years ago	<0–20)	<20–40)	<40–60)	<60–80)	<80–100>
4	The level of implementation of intelligent solutions in the field of city logistics	0—The solution is not planned for implementation 1—The solution is planned to be implemented in the next 5–10 years 2—The solution is planned to be implemented in the perspective of 1–5 years 3—The solution is being implemented 4—The solution is implemented	<0–20)	<20–40)	<40–60)	<60–80)	<80–100>

Source: own work

Methodology for Assessing the Impact of the City's Logistics Maturity on the City's Advancement Level in Terms of Smart City Solutions

As part of the research carried out in this monograph, an analysis of the relationship between the advancement of Smart City solutions and the level of logistics maturity of the studied cities was also carried out. The aim of this part of the research was to answer the following research questions: Does the level of development of a Smart City influence its logistics maturity and how?

The stages of this analysis, together with the list of statistical tools used to carry them out, are included in Table 5.5.

The results of the analysis carried out in this way are diagnostic and allow to assess to what extent the implementation of the Smart City concept by the authorities of Polish cities affects their logistics maturity. In addition, thanks to the identification of key determinants of logistics maturity, they enable the setting of directions and actions to improve the functioning of city logistics. Research in this area has not been conducted so far, and its results can be used by both municipal authorities and other stakeholders involved in the development of logistics infrastructure.

Table 5.5 Research stages and tools used in the process of assessing the relationship between the advancement of Smart City solutions and the level of maturity

No.	Analysis stage	Research tool
1	Preparation of descriptive statistics for both assessed aspects to conduct a preliminary comparative analysis	Parameters of descriptive statistics for the final and partial grades (criteria for assessing maturity from 1 to 4): • Arithmetic mean • Median, lower, and upper quartiles • Standard deviation and coefficient of variation • Kurtosis and skewness of distribution
2	Classification of the examined cities, taking into account both assessed aspects, aimed at determining the structure and frequency of cities in particular groups	Describing cities with the use of both examined aspects: the advancement of Smart City solutions and the level of logistics maturity. Determining the frequency and structure of the cities studied, taking into account the following divisions: • Logistics maturity 1; advancement of Smart City solutions from 0 to 5 • Logistics maturity 2; advancement of Smart City solutions from 0 to 5 • Logistics maturity 3; advancement of Smart City solutions from 0 to 5 • Logistics maturity 4; advancement of Smart City solutions from 0 to 5 • Logistics maturity 5; advancement of Smart City solutions from 0 to 5
3	Assessment of the correlation between the level of advancement of intelligent solutions and the level of logistics maturity of the cities studied, allowing to identify the strength and direction of the relationship between the indicated variables	Calculation of the Pearson linear correlation coefficient that allows to assess the strength and direction of the linear relationship between the studied variables. The value of this coefficient ranges from <-1.00; 1.00>. The value of 0.00 means there is no linear relationship between the variables. On the other hand, the closer the value is to 1.00 or -1.00, the stronger the linear relationship The research identified a correlation between the advancement of Smart City solutions and the level of logistics maturity of the cities studied. The significance level of 0.05 was adopted in the analysis

(continued)

Table 5.5 (continued)

No.	Analysis stage	Research tool
4	Identification of the impact of individual criteria of logistics maturity (from 1 to 4) and the advancement of Smart City solutions on the final level of maturity of the cities studied	Parameterising a multiple regression function that allows you to estimate the influence of multiple independent variables (X_1, X_2, X_3, ..., X_n) on the dependent variables (Y). This function takes the following form: $$Y = b_0 + b_1 \times x_1 + b_2 \times x_2 + ... + b_n \times x_n + \varepsilon$$ Where: Y – Dependent variable explained by the function, x_1, x_2, ..., x_n – Independent, explanatory variables, b_1, b_2, b_3, ..., b_n – Parameters defining the contribution of individual independent variables to the explanation of the model, – random component rest of the model. The multiple regression model is verified using: (a) Evaluation of the statistical significance of individual variables in the model (the analysis adopted a significance level of 0.05) (b) Multiple determination coefficient (R^2), which is a measure of the fit of the model with values in the range <0.00; 1.00>, where the value 1.00 means perfect matches and 0.00 no fit (c) The corrected multiple determination coefficient, which, similarly to (R^2), allows to assess the fit of the model, but is insensitive to the number of variables and the sample size, takes values in the range <0.00; 1.00>, where the value 1.00 means perfect matches and 0.00 no match

Source: own work

References

Calzada, I. (2017). The techno-politics of data and smart devolution in city-regions: Comparing Glasgow, Bristol, Barcelona, and Bilbao. *Systems, 5*, 18. https://doi.org/10.3390/systems5010018

Fernandez-Anez, V., Fernandez-Guell, J. M., & Giffinger, R. (2018). Smart city implementation and discourses: An integrated conceptual model. The case of Vienna. *Cities, 78*, 4–16.

Girardi, P., & Temporelli, A. (2017). Smartainability: A methodology for assessing the sustainability of the smart city. *Energy Procedia, 2017*(111), 810–816.

Jonek-Kowalska, I., & Wolniak, R. (2021). Economic opportunities for creating smart cities in Poland. Does wealth matter? *Cities, 14*, 1–16. art. no. 103222.

Lee, J. H., Hancock, M. G., & Hu, M.-C. (2014). Towards an effective framework for building smart cities: Lessons from Seoul and San Francisco. *Technological Forecasting and Social Change, 89*, 80–99.

McAdam, M., & Debackere, K. (2018). Beyond 'Triple Helix' toward 'Quandruple Helix' models in regional innovation systems: Implications for theory and practice. *R&D Management, 48*, 3–6.

Schütz, F., Heidingsfelder, M., & Schraudner, M. (2019). Co-shaping the future in quadruple helix innovation systems: Uncovering public preferences toward participatory research and innovation. *Journal of Design, Economics, and Innovation, 5*, 128–146.

Taratori, R., Rodriguez-Fiscal, P., Pacho, M. A., Koutra, S., Pareja-Eastaway, M., & Thomas, D. (2021). Unveiling the evolution of innovation ecosystems: An analysis of triple, quadruple, and quintuple helix model innovation systems in European case studies. *Sustainability, 13*, 7582. https://doi.org/10.3390/su13147582

Wolniak, R., & Jonek-Kowalska, I. (2021). The level of the quality of life in the city and its monitoring. *Innovation (Abingdon), 4*(3), 376–398.

6

Logistics Maturity of Polish Cities on the Way to Smart City

The Level of Development of Polish Cities in the Implementation of the Smart City Concept: Research Results

The segregation of the cities participating in the study, carried out in accordance with the rules described in the previous chapter, made it possible to assess the level of their advancement in the implementation of the Smart City concept. The results in terms of the level achieved are shown in Fig. 6.1.

The largest number of Polish cities participating in the survey represented levels 0 and 1 in terms of the implementation of the SC concept. This means that 96 cities do not know what the characteristics of Smart City are and do not plan to implement Smart City solutions, and 102 cities know the concept itself but are not interested in using it. It can therefore be concluded that the knowledge and interest of SC in Poland is low, because only 30% of the surveyed units predict a quick development of the Smart City concept and the achievement of subsequent stages of advancement in this development.

© The Author(s), under exclusive license to Springer Nature Switzerland AG 2022
M. Kramarz et al., *Urban Logistics in a Digital World*,
https://doi.org/10.1007/978-3-031-12891-2_6

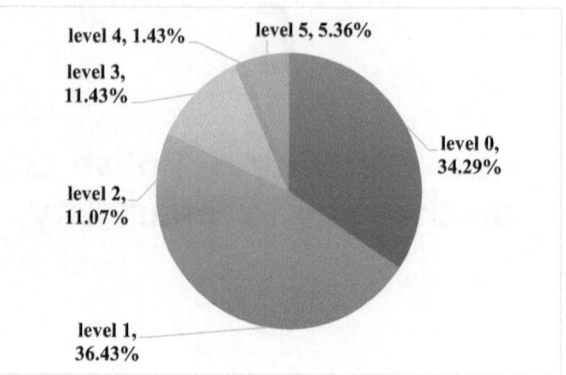

Fig. 6.1 Assessment of the advancement level in the implementation of the Smart City concept: survey results [level; %]. Source: own work based on the results of the survey

The second and third levels of advancement in the implementation of the Smart City concept were represented by a similar number of cities, amounting to 31 and 32, respectively, so 63 cities want to include the assumptions for creating SC in their city strategy, and 32 of them additionally carry out a comprehensive evaluation of their activities in the context of "being smart". The declarations made by these cities indicate a real interest in the dynamic development and improvement of the quality of life of local communities.

Interestingly, the fourth level of advancement in the implementation of the SC concept was represented by only 4 units, which is less than 1.5% of the studied sample. In view of the above, cities do not see the need to cooperate with the science sector in the process of planning and implementing Smart City solutions. Therefore, they do not want to implement the principles defined within the triple helix, meaning a partnership on the line: city–business–science. This is a disturbing conclusion that requires diagnostic research to address the causes of such reluctance. They may result from both ignorance of the scientific offer and bad experiences resulting from the cooperation of the city authorities with the scientific community so far. It should also be added that the lack or a small scope of this cooperation may pose a serious threat to the transfer of knowledge and modern technologies to cities and reduce the level of their innovation, and thus hinder the development of the Smart City concept. Therefore, strengthening the cooperation between cities and science should be considered one of the key recommendations resulting from this research.

The fifth level of advancement in the implementation of the Smart City concept was characteristic of the 15 surveyed cities, which not only knew the principles of creating Smart City solutions and were interested in them, but also included them in the scope of strategic activities and implemented them in cooperation with the science sector and local communities. In these cities, there is a quadruple helix covering cooperation in the system: city–business–science–local community, promising for the development of Smart City and improving the quality of life of the inhabitants.

In the further part of the subchapter, an analysis of the demographic and economic conditions characterising the cities classified with the above-described levels of advancement in the implementation of SC solutions was carried out. For this purpose, record questions referring to the number of inhabitants living in a given city and budget income per capita and reflecting the level of wealth of the examined city were used. In the first criterion, the respondents could choose the following answer options:

- up to 5000 inhabitants,
- from 5001 to 10,000 inhabitants,
- from 10,001 to 25,0000 inhabitants,
- from 25,001 to 50,000 inhabitants,
- from 50,001 to 100,000 inhabitants,
- above 100,000 inhabitants.

In the case of the second criterion, the cafeteria included the following income ranges:

- below PLN 1000,
- from PLN 1001 to PLN 2000,
- from PLN 2001 to PLN 3000,
- from PLN 3001 to PLN 4000,

- from PLN 4001 to PLN 5000,
- above PLN 5000.

The results of the demographic and income analysis for the highest fifth level are presented in Figs. 6.2 and 6.3.

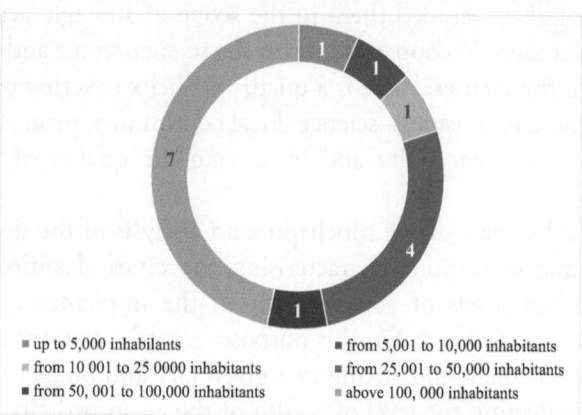

- up to 5,000 inhabilants
- from 10 001 to 25 0000 inhabitants
- from 50, 001 to 100,000 inhabitants
- from 5,001 to 10,000 inhabitants
- from 25,001 to 50,000 inhabitants
- above 100, 000 inhabitants

Fig. 6.2 Cities with level 5 taking into account the number of inhabitants. Source: own work based on the results of the survey

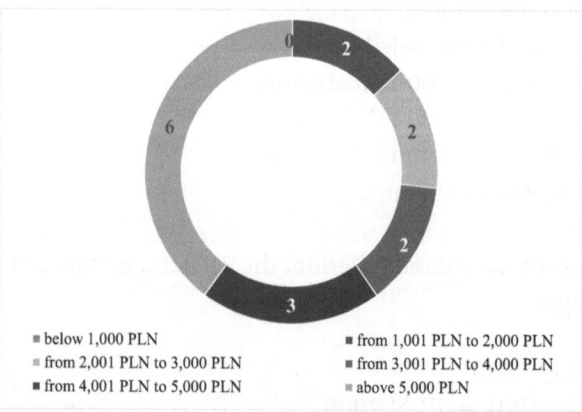

- below 1,000 PLN
- from 2,001 PLN to 3,000 PLN
- from 4,001 PLN to 5,000 PLN
- from 1,001 PLN to 2,000 PLN
- from 3,001 PLN to 4,000 PLN
- above 5,000 PLN

Fig. 6.3 Cities with level 5, taking into account budget revenues per capita. Source: own work based on the results of the survey

The obtained data show that the highest advancement level was represented by large and very large cities. In most of them, the number of inhabitants exceeded 25,000, and in as many as 7–100,000. They were also cities with above-average budget revenues. In most of them, the per capita income ratio exceeded PLN 3000. These results confirm previous observations and experiences, which indicate that the Smart City concept is characteristic and available for large cities with good and very good material status. Nevertheless, it is worth adding that the analysed group also included three to four small and less affluent cities, which means that the stereotypical perception of smart cities can be changed and it is not a limiting or excluding concept.

The results of the demographic and income analysis for the fourth level of advancement in the implementation of the Smart City concept are shown in Figs. 6.4 and 6.5. They show that—this rather high level is also represented by mostly very large cities, with the highest level of income per capita, which confirms the observations described above.

At the third level of advancement in the implementation of the Smart City concept, a much greater demographic and income differentiation can be observed, which is illustrated in Figs. 6.6 and 6.7. Although this list does not include the smallest and poorest cities, the number of units in the remaining analysed categories is fairly balanced.

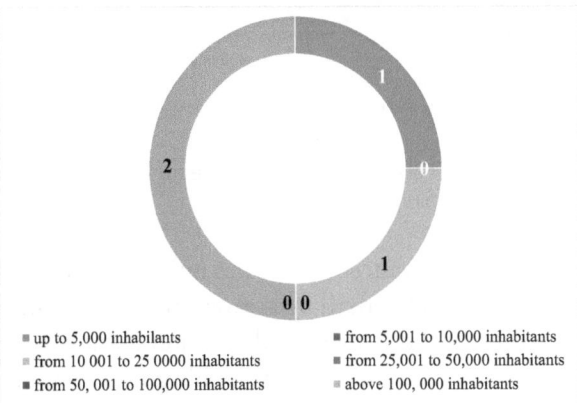

■ up to 5,000 inhabilants   ■ from 5,001 to 10,000 inhabitants
▨ from 10 001 to 25 0000 inhabitants   ■ from 25,001 to 50,000 inhabitants
■ from 50, 001 to 100,000 inhabitants ▨ above 100, 000 inhabitants

Fig. 6.4 Cities with level 4 taking into account the number of inhabitants. Source: own work based on the results of the survey

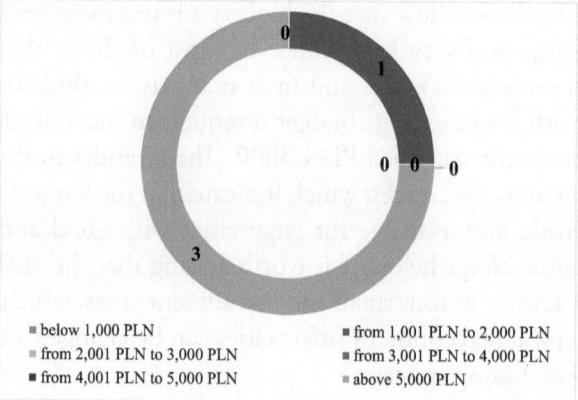

Fig. 6.5 Cities with level 4, taking into account budget revenues per capita. Source: own work based on the results of the survey

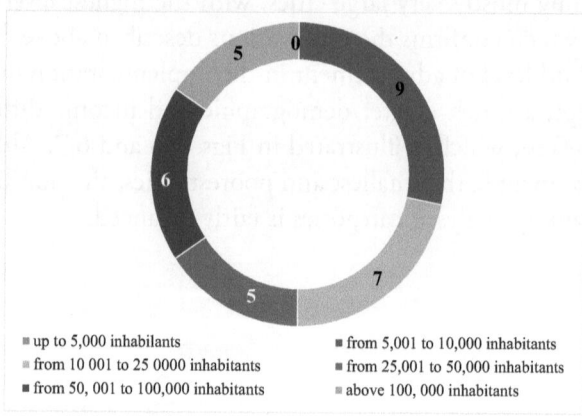

Fig. 6.6 Cities with level 3 taking into account the number of inhabitants. Source: own work based on the results of the survey

The second level of advancement in the implementation of the Smart City concept is also characterised by greater income diversity. It is dominated by medium-sized cities with a population of 5–25,000. Interestingly—in the economic context—this group includes both cities with a relatively low budget income per capita—not exceeding PLN 2000, and cities with a higher level of affluence, where the budget income per capita is in the range of 3000–5000 PLN (Figs. 6.8 and 6.9).

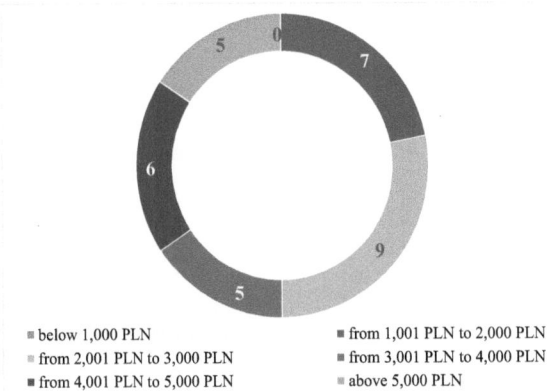

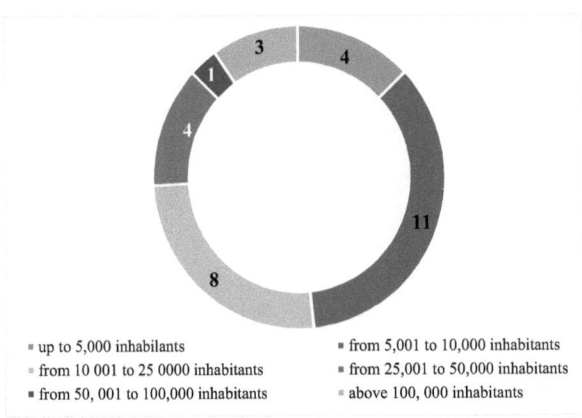

Fig. 6.7 Cities with level 3, taking into account budget revenues per capita. Source: own work based on the results of the survey

Fig. 6.8 Cities with level 2 taking into account the number of inhabitants. Source: own work based on the results of the survey

In the case of the last two levels, defined as 1 and 0, and covering the largest number of cities studied, two regularities can be observed. The first is related to the growing share of small and medium-sized cities with a population not exceeding 25,000. The second one relates to the decreasing level of budget revenues per capita, which in most of the surveyed entities rarely exceed PLN 3000. Detailed data in this regard is presented in Figs. 6.10, 6.11, 6.12, and 6.13.

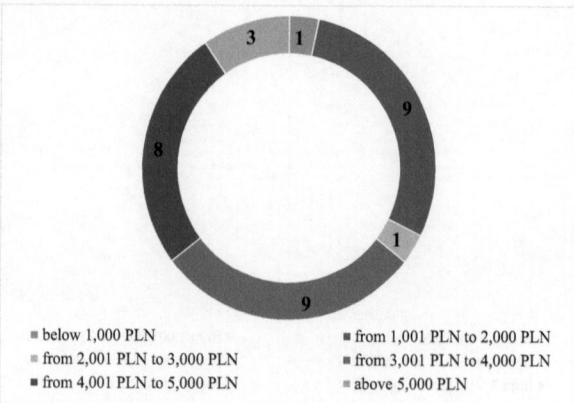

Fig. 6.9 Cities with level 2, taking into account budget revenues per capita. Source: own work based on the results of the survey

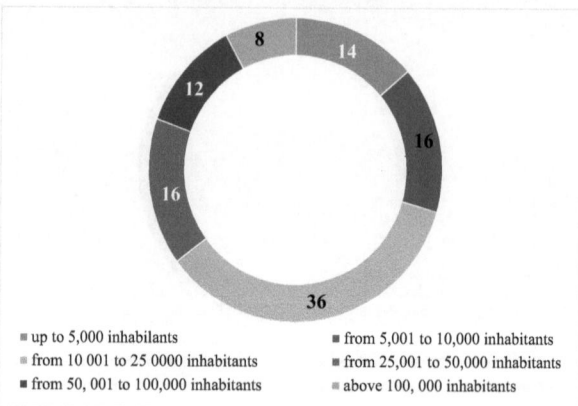

Fig. 6.10 Cities with level 1 taking into account the number of inhabitants. Source: own work based on the results of the survey

And so, the first level of advancement in the implementation of the Smart City concept is mainly represented by cities with a population of no more than 25,000 inhabitants, and in which the level of budget revenue does not exceed PLN 3000. In these cities, the city authorities are familiar with the Smart City concept, but are not interested in its implementation.

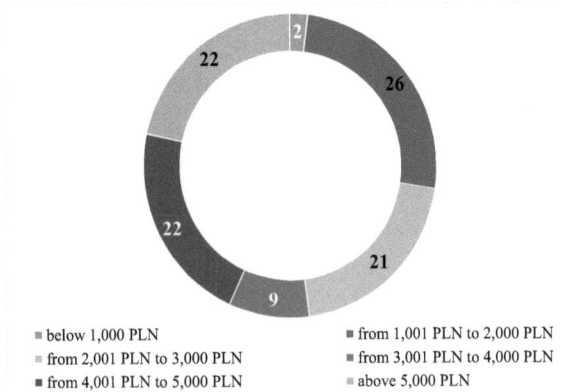

Fig. 6.11 Cities with level 1, taking into account budget revenues per capita. Source: own work based on the results of the survey

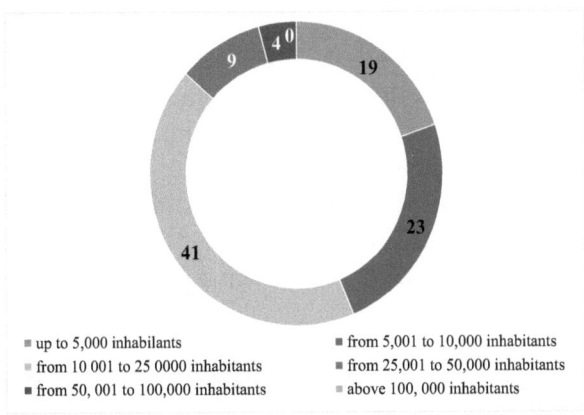

Fig. 6.12 Cities with level 0 taking into account the number of inhabitants. Source: own work based on the results of the survey

In the case of cities at level 0, that is, those who do not know and are not interested in implementing the Smart City concept, the number of large and very large cities is significantly decreasing (there are no cities inhabited by more than 100,000 inhabitants, and the number of cities inhabited by 25,000–100,000 is only 13, which is less than 14% of the respondents). The number of cities where the per capita income does not

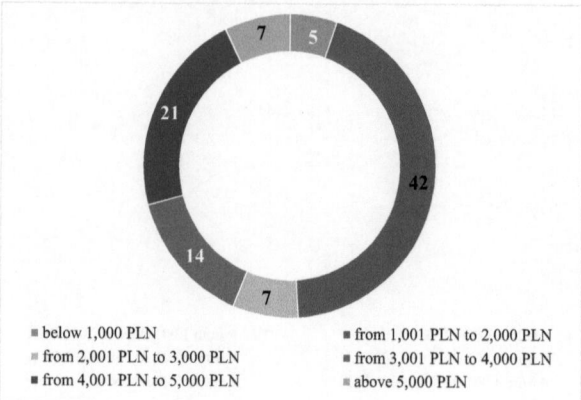

Fig. 6.13 Cities with level 0, taking into account budget revenues per capita. Source: own work based on the results of the survey

exceed PLN 3000 is also growing. This confirms the conclusions drawn in the case of the analysis of cities representing the highest levels and states that the level of maturity in the implementation of the Smart City concept increases with the increase in the number of inhabitants and city budget revenues in terms of per capita.

In the further part of the study, the levels of advancement of Polish cities in the implementation of the Smart City concept, identified in this chapter, will be confronted with their logistics maturity.

The Level of Logistics Maturity of the Cities Studied: Research Results

The maturity levels determined on the basis of the criteria distinguished at the stage of literature research made it possible to analyse what characterises cities at a given maturity level. It should be emphasised once again that the transition to the next level of maturity requires meeting all criteria at this level. It is an essential assumption made in the research, which affects the results obtained.

By comparing the levels of logistics maturity of the cities studied (Figs. 6.14 and 6.15), a picture of the current advancement of logistics solutions in Poland was obtained. First of all, there was a clear advantage

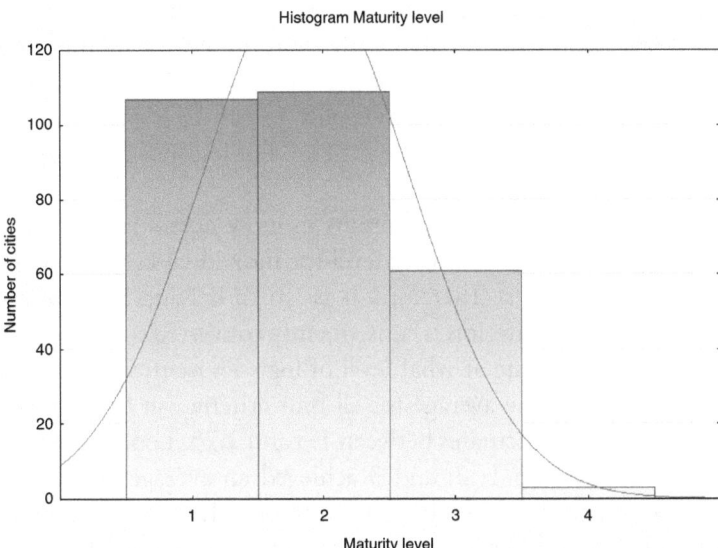

Fig. 6.14 Histogram of logistics maturity of the studied cities. Source: own work

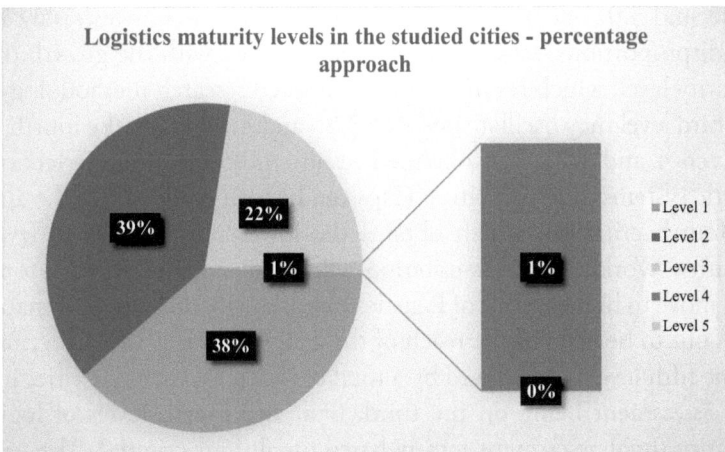

Fig. 6.15 The level of logistics maturity of Polish cities—percentage approach. Source: own work

of cities at the first and second maturity levels (107 cities—38% and 109 cities—39%, respectively). Cities with the third level of maturity constitute a smaller group, because 22% (61 cities). Only three cities reached the fourth level of maturity, which is only 1% of the 280 cities surveyed. There was no representative of the fifth level of logistics maturity among the examined cities.

When analysing the obtained results in more detail, it can be noticed that cities with the first maturity level for individual criteria sometimes achieve extreme results. Therefore, it is worth delving into these data. Assuming that each criterion is consequently converted to a 5-point scale, which allows to indicate at what level of logistics maturity a given criterion is of each city, the average for all four criteria can be indicated. At level 1, this average fluctuates between 1.5 and 2.75. Looking at the individual criteria, criteria 1, 2, and 3 achieved an average above 2 (2.03, 2.27, 2.29, respectively), while criterion 4 only 1.45. It was mainly this criterion that determined the cities to remain at the first level of maturity. Similar conclusions can be drawn when analysing the second level of logistics maturity (the average in cities oscillates between 2.25 and 4.0). On the other hand, among the criteria, the first has an average of 2.59, the second 3.01, the third 2.90, and the fourth is the lowest, only 2.18. The disproportions between the criteria decrease with the growth of the maturity level, which results from the adopted research methodology (on the third level they oscillate between 3.25 and 4, while on the fourth level between 4 and 4.5). The obtained results indicate the imperfection of logistic systems of Polish cities. Gaps can be seen especially in the area of the fourth criterion, which obtains the lowest results at each level of maturity. Working on this criterion will allow many cities to automatically move to higher levels of logistics maturity. The fifth level of maturity turns out to be beyond the reach of the analysed cities. The first criterion on the fifth level was assessed by a total of six cities, with five cities in the final assessment being on the third, first, and fourth levels of logistics maturity (final assessment jointly based on all four criteria). The second criterion reached the fifth level in a total of 17 cities, wherein this criterion was assessed on the fifth level by 6 cities which, as a result, were placed on the second level, and 11 cities which were placed on the third level in the final assessment. In the case of criterion 3, the fifth level was obtained by a total of 18 cities, including 2 from the first level in the final maturity assessment, 6 from the second level, 9 from the third level, and 1 from the fourth level of maturity. The fourth criterion was assessed on the fifth level six times (including one city which, as a result, reached the second maturity level, four cities which are generally assessed at the third maturity level, and one city at the fourth maturity level).

Another element of the research was drawing attention to logistics maturity in geographical terms—in the context of individual regions of the country (voivodships). The conducted analyses of the logistics maturity of Polish cities broken down into voivodeships (Figs. 6.16 and 6.17) did not show any significant differences. The analysis of the results allows, first of all, to determine that none of the voivodships, and thus none of the surveyed cities, has reached the highest level of logistics maturity. When analysing the average level of maturity, the first place is taken by the Łódzkie voivodeship (2.25). A relatively high level of maturity was also recorded in the following voivodships: Mazowieckie (2.08), Śląskie (2.09), Zachodniopomorskie (2.08), and Świętokrzyskie (2.0). The average in the remaining voivodships did not exceed the level of 2. The average maturity levels for all voivodeships are in the range of 1.57–2.25,

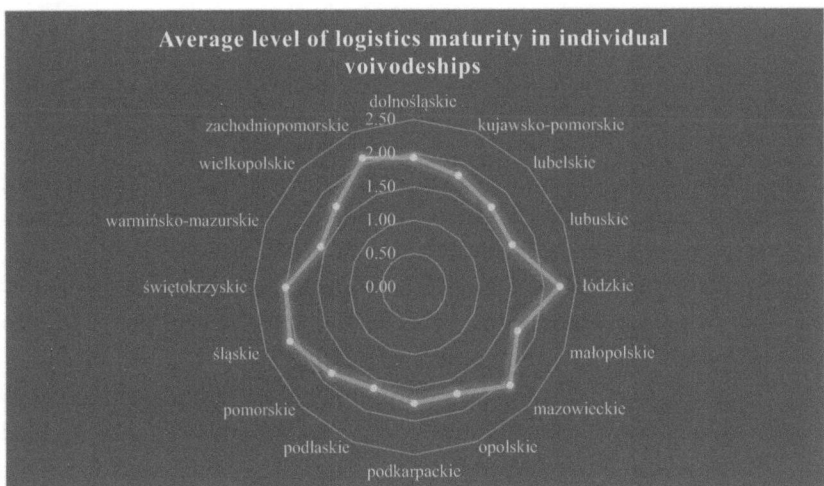

Fig. 6.16 Average level of logistics maturity in individual voivodeships. Source: own work

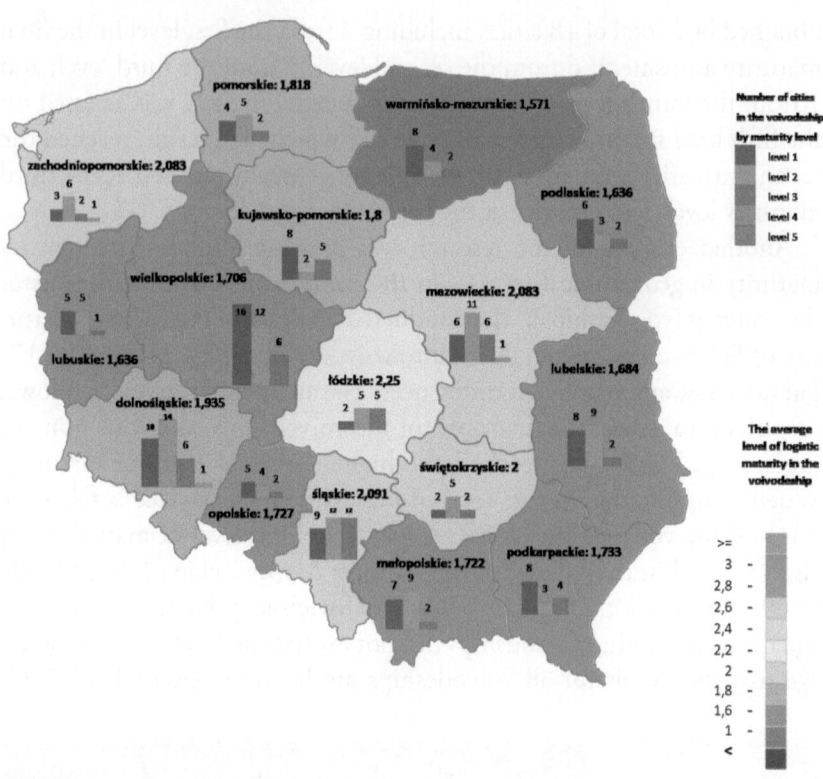

Fig. 6.17 Map of Poland with the levels of logistics maturity of individual voivodeships. Source: own work

which should be considered a small range of results. The highest level of logistics maturity is 2.25 (Łódzkie Voivodeship), four voivodships are in the range from 2.0 to 2.2: Śląskie, Mazowieckie, Zachodniopomorskie, and Świętokrzyskie. The remaining 11 voivodships were below the maturity level 2. Geographical location is therefore not a factor that strengthens or inhibits the development of city logistics.

Average values for voivodships do not give a full picture of the structure of the maturity levels of cities in these voivodships. A more detailed picture was obtained by presenting the distribution of maturity levels for individual provinces (Fig. 6.18).

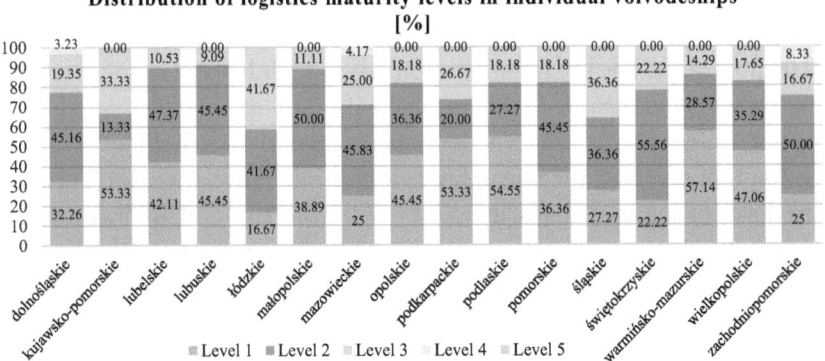

Fig. 6.18 The structure of logistics maturity levels in individual provinces [%]. Source: own work

Cities that are currently at the fourth level of maturity can be an example of good practice and become a benchmark for other cities, especially those with similar characteristics. Such cities, among all respondents, are in Dolnośląskie, Mazowieckie, and Zachodniopomorskie voivodships. The Łódzkie, Śląskie, and Świętokrzyskie voivodships do not have representatives at the fourth level of maturity among the examined cities. The higher average value obtained by these voivodships is the result of a large share of cities at the third level of maturity and a small percentage of cities at the first level of logistics maturity. In the Łódzkie voivodship, cities at the third level of maturity account for 41.67% of the examined cities. It is the best result compared to all voivodships. For comparison, among the cities of the Kujawsko-Pomorskie Voivodeship, those at the first maturity level dominate (they constitute as much as 53.33% of the examined cities), similarly to the Podkarpackie Voivodeship. On the other hand, in Podlaskie voivodship, cities at the first level of maturity are as high as 54.55%, and in Warmińsko-Mazurskie—57.14%. Among the examined cities, the most surprising result is the distribution of maturity levels in the Śląskie voivodeship—the first maturity level represents 27.27% with a slightly higher representation of the second and third levels (36.36%). It seems that the Górnośląska-Zagłębiowska Metropolis has introduced top-down regulations regarding cities, which should significantly reduce the disproportions between the development of city logistics and the city's approach to implementing innovative solutions in this area. The Metropolitan Transport System, including the strategic approach to transport at the level of the metropolis and individual cities, focuses only on passenger transport; however, the solutions in this area are significantly unified also at the level of cooperation with stakeholders.

Another aspect of the research was to indicate the relationship between the level of logistics maturity achieved by the city and its income per capita. Six income ranges were used in the research. In each of these groups, the average level of maturity achieved by the cities assigned to a given group was determined (Fig. 6.19).

As shown in the chart (Fig. 6.19), it cannot be concluded that the average level of maturity increases with the increase in the range of the city's income per capita. The dominant feature for the first and second income ranges represents the first level of logistics maturity. In the case of other income ranges, the dominant is identical and is described by the second level of logistics maturity. It is clearly visible that in the lowest per capita income range (below PLN 1000), the average maturity level of the cities assigned to it is the lowest (1.38). At the same time, the highest average level of maturity (2.76) is visible in the case of the highest range of per capita income (over PLN 5000). It can also be noticed that the result for this range (2.76) definitely differs in size from all other maturity levels

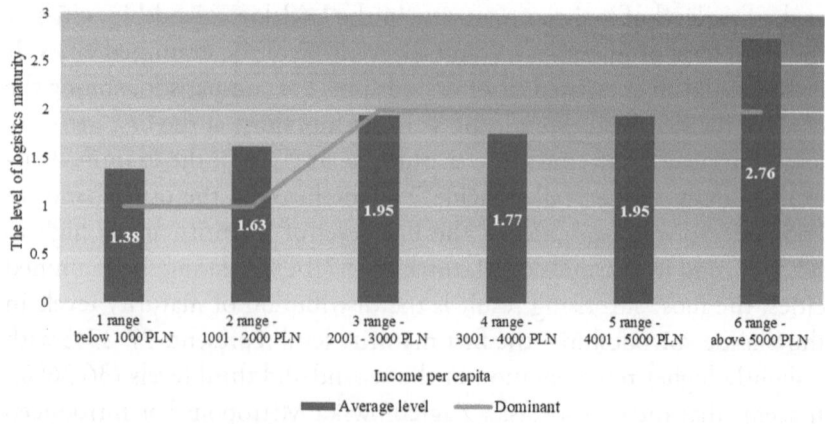

Fig. 6.19 Average level of logistics maturity in individual ranges of income per capita. Source: own work

obtained in other income ranges. For the remaining ranges, it is difficult to speak of a clear tendency. The difference between the maturity in the sixth (highest) income range and the highest maturity in the remaining income range is 0.81. The differences between maturities for the remaining ranges are slight. It should be noted that between the highest maturity level of the first five income ranges (maturity level 1.95 for the third and fifth income ranges) and the lowest maturity level (1.38 for the first income range) the difference is only 0.57.

To confirm the initially formulated conclusions regarding the relation between the amount of income per capita and the level of logistics maturity, an analysis of *Pearson's r correlation* was performed. The analysis of the correlation between the level of income per capita and the level of logistics maturity of the city showed (Fig. 6.20) that there is no statistically significant relationship between them ($r = 0.10300$).

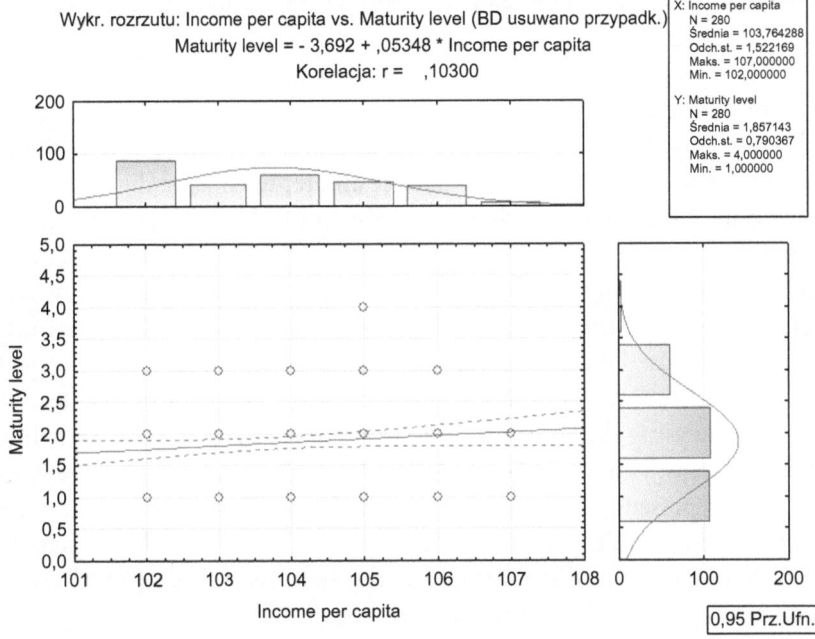

Fig. 6.20 Correlation between the amount of income of cities per capita and their level of logistics maturity. Source: own work

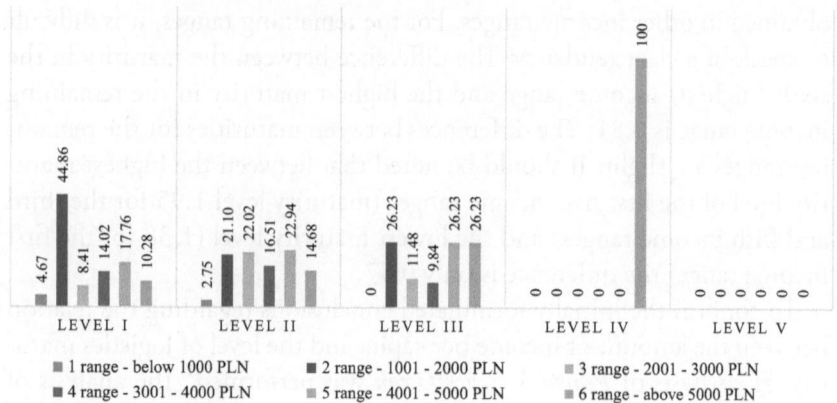

Fig. 6.21 Distribution of cities in terms of income per capita for individual levels of logistics maturity. Source: own work

Figure 6.21 shows the lack of a strong and unambiguous relationship between the amount of income of a city per capita and the level of maturity achieved.

When analysing the distribution of cities in terms of income per capita, only in the case of the fourth level of maturity, one income range is dominant—all cities obtained the highest level of income. However, this result is difficult to treat as fully reliable and representative due to the fact that only three cities were classified at the fourth level of maturity. This is too little representation to clearly confirm the conclusion. The first range of income (the lowest) is represented by only 8 cities out of 280 surveyed. Hence, it is not surprising that the share of these cities at each maturity level is low (at the first and second levels it is 4.67 and 2.75%, respectively, and there are no cities of this type at the third and fourth levels). When analysing the distribution of cities at the first, second, and third levels of logistics maturity, only in the case of the first level is a large share of cities representing one income group visible. This is the second range of income (from PLN 1001 to PLN 2000 per capita—44.86% of cities). The remaining city distributions, at the first, second, and third levels, constitute a relatively comparable structure.

Another factor that was taken into account in the search for determinants of the logistics maturity of cities was the size of the city expressed

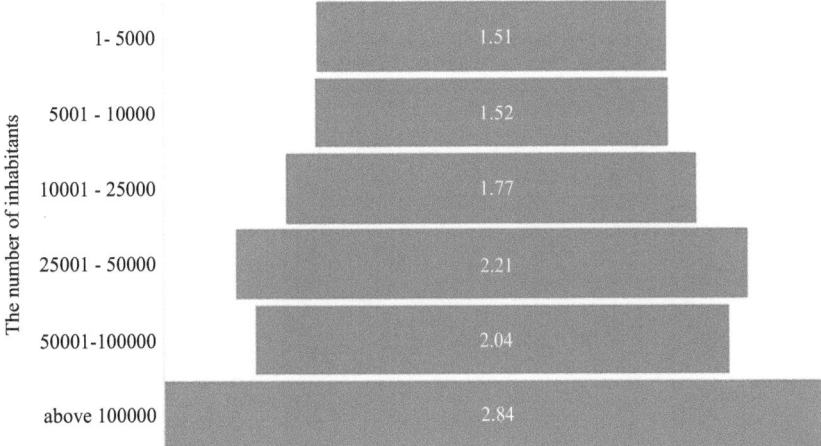

Fig. 6.22 Average level of logistics maturity of cities in individual population groups. Source: own work

by the number of its inhabitants. Thus, in the first step, the average level of logistics maturity achieved by cities in the six assumed population ranges was analysed (Fig. 6.22). The results show an increase in the level of logistics maturity in line with population growth. The only exception is cities in the group from 50,001 to 100,000 inhabitants, which recorded a slightly lower average value than cities with a population of 25,001 to 50,000.

As in the case of the geographical location criterion (voivodships), the differentiation of logistics maturity in individual population groups is small and oscillates between 1.52 and 2.84 of the average maturity level.

To confirm the observed relationship between the city size and the level of logistics maturity, *the Pearson's r* correlation analysis was performed. It showed that there is a significant, positive, and moderately strong relationship between the city size and its level of logistics maturity ($r = 0.45069$). Thus, it was confirmed that with the increase in the number of inhabitants in the city, its level of logistics maturity increases (Fig. 6.23).

Moreover, the correlation analysis showed (Fig. 6.24) that level 1 of logistics maturity was characteristic of cities with a population of 5001 to

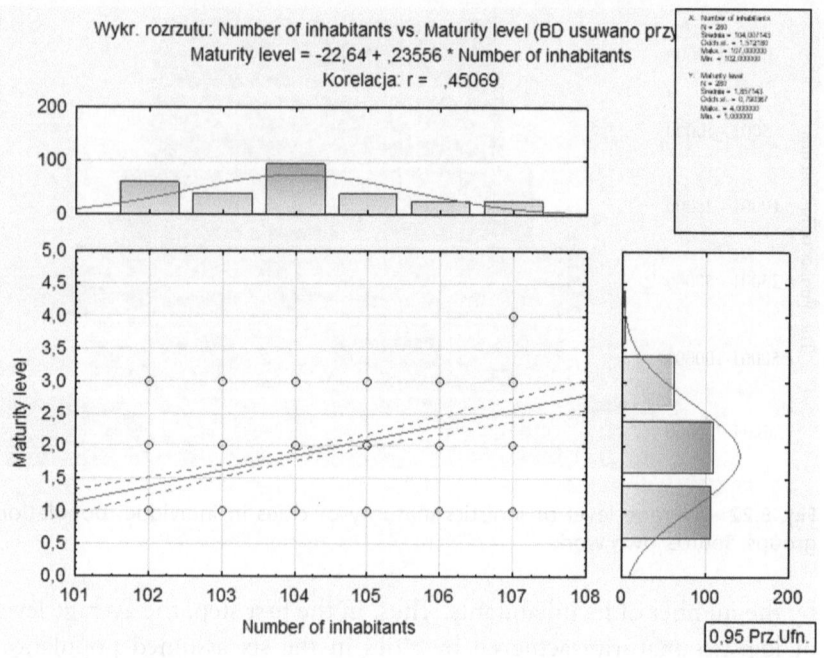

Fig. 6.23 Correlation between the size of cities and their level of logistics maturity. Source: own work

25,000, while level 2 of cities from 10,000 to 50,000 inhabitants, level 3 of cities from 10,000 to 100,000 inhabitants, while only cities with more than 100,000 inhabitants have reached level 4 in terms of logistics maturity.

Additionally, in relation to the voivodships with the highest level of logistics maturity, an analysis of the correlation between the size of cities in the voivodship and the amount of income per capita and the level of logistics maturity of the voivodship cities was carried out.

Regarding the analysed correlations for Łódzkie voivodship, there is a relationship between the analysed variables, which, however, shows a low correlation. In the case of the relationship between the number of inhabitants and the level of logistics maturity of the city, the correlation is $r = 0.25749$, which means that with the increase in the number of

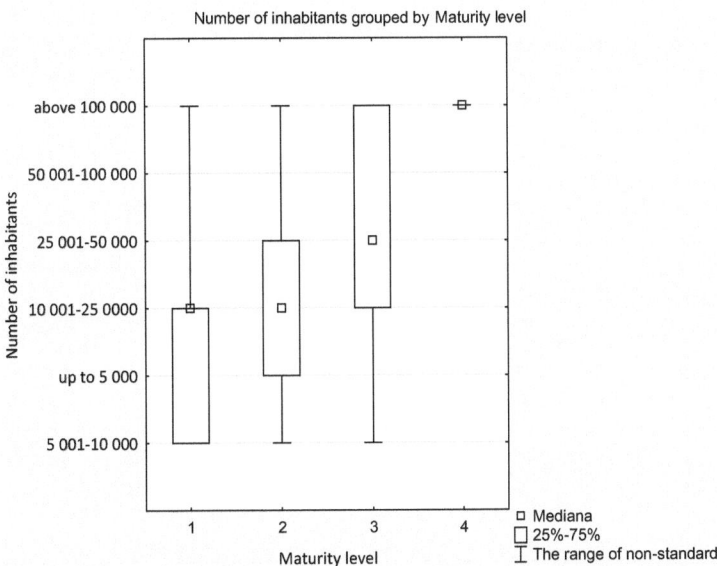

Fig. 6.24 A box plot reflecting the clustering of city sizes according to their level of logistics maturity. Source: own work

inhabitants in a voivodeship, its level of logistics maturity increases (Fig. 6.25), and in relation to the analysis between the amount of income per capita and the level of logistics maturity there is a negative correlation ($r = -0.2705$), which means that an increase in the amount of income per capita is accompanied by a decrease in the level of logistics maturity (Fig. 6.26).

Regarding the analysed correlations for Śląskie voivodship, there is a relationship between the analysed variables, which, however, shows a low correlation. In the case of the relationship between the number of inhabitants and the level of logistics maturity of the city, the correlation is $r = 0.26900$, which means that with the increase in the number of inhabitants in a voivodeship, its level of logistics maturity increases (Fig. 6.27). The analysis of the correlation between the level of income per capita and the level of logistics maturity of the city showed (Fig. 6.28) that there is no statistically significant relationship between them ($r = 0.0168$).

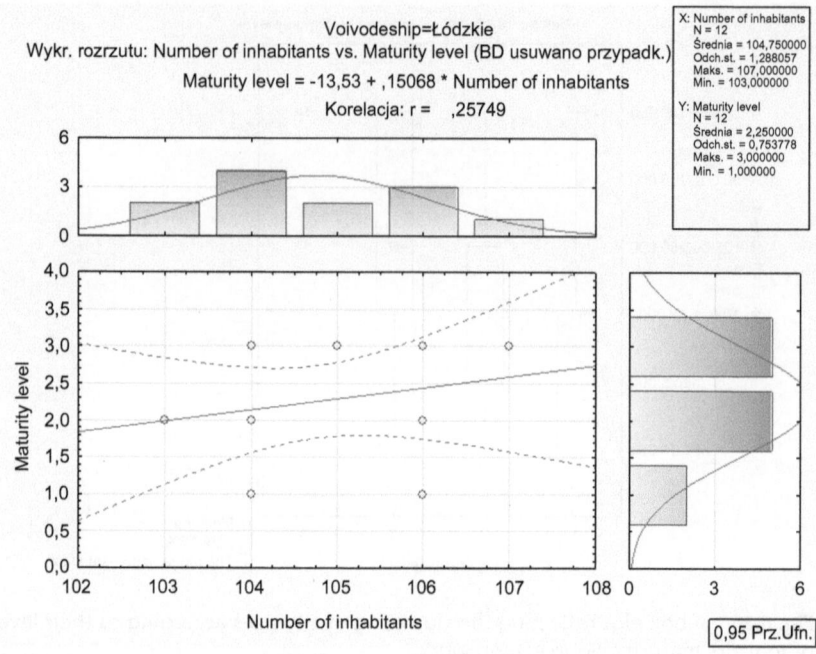

Fig. 6.25 Correlation between the size of cities in Łódzkie voivodeship and their level of logistics maturity. Source: own work

Regarding the analysed correlations for Mazowieckie voivodship, there is a relationship between the analysed variables, which shows a high correlation. In the case of the relationship between the number of inhabitants and the level of logistics maturity of the city, the correlation is $r = 0.66012$, which means that with the increase in the number of inhabitants in a voivodeship, its level of logistics maturity increases (Fig. 6.29).

Regarding the analysis, there is a moderate correlation between the number of incomes per capita and the level of logistics maturity ($r = 45,254$), which means that with the increase in the amount of income per capita, the level of logistics maturity increases (Fig. 6.30).

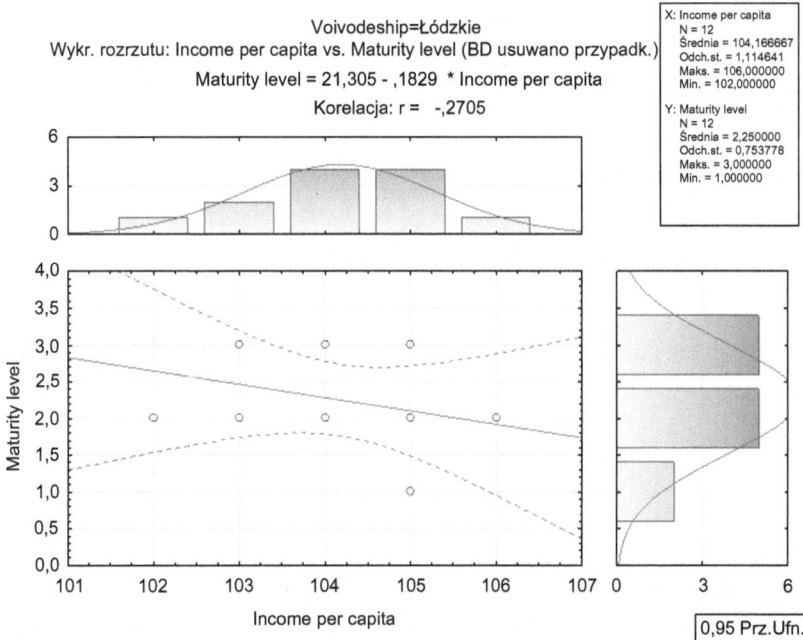

Fig. 6.26 Correlation between the amount of city income per capita in Łódzkie voivodeship and its level of logistics maturity. Source: own work

Assessing the Impact of the City's Logistics Maturity on the City's Advancement Level in Terms of Smart City Concept

This monograph confronted the assessment of Smart City solutions advancement with the level of logistics maturity of the surveyed cities. The results of this comparison are presented taking into account the following research stages and goals:

- preparation of descriptive statistics for both assessed aspects to conduct a preliminary comparative analysis,
- classification of the examined cities, taking into account both assessed aspects, aimed at determining the structure and frequency of cities in particular groups,

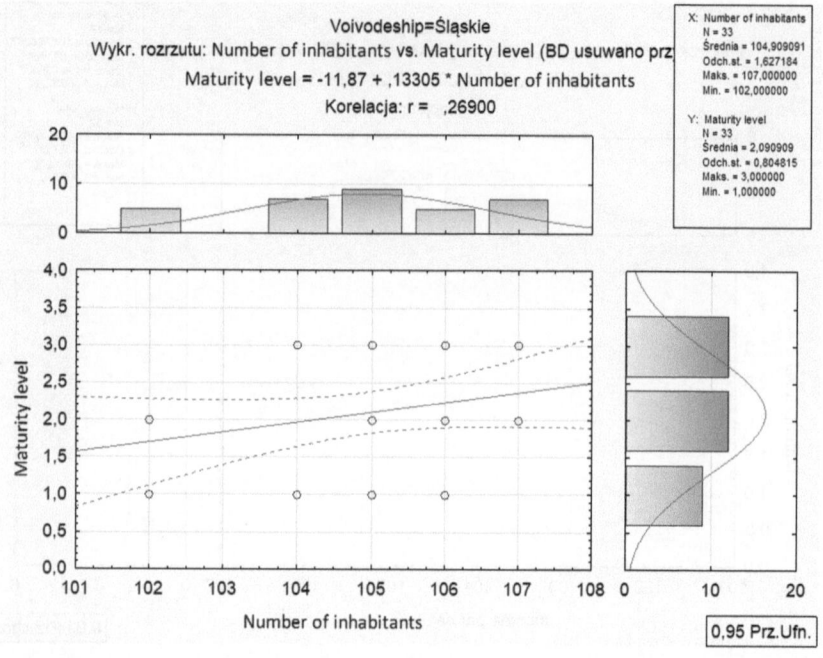

Fig. 6.27 Correlation between the size of cities in Śląskie Voivodeship and their level of logistics maturity. Source: own work

- assessment of the correlation between the level of advancement of intelligent solutions and the level of logistics maturity of the cities studied, allowing to identify the strength and direction of the relationship between the indicated variables,
- identification of the impact of individual criteria of logistics maturity (from 1 to 4) and the advancement of Smart City solutions on the final level of maturity of the cities studied.

Thus, Table 6.1 shows the results of the first step mentioned above.

The data in Table 6.1 show that the studied cities were characterised by a very low level of advancement of Smart City solutions and a slightly higher, but still low, level of logistics maturity. Half of the surveyed units did not exceed the score 1.00 for the Smart City criterion and the score 2.00 for the logistics maturity criterion. Therefore, Polish cities are

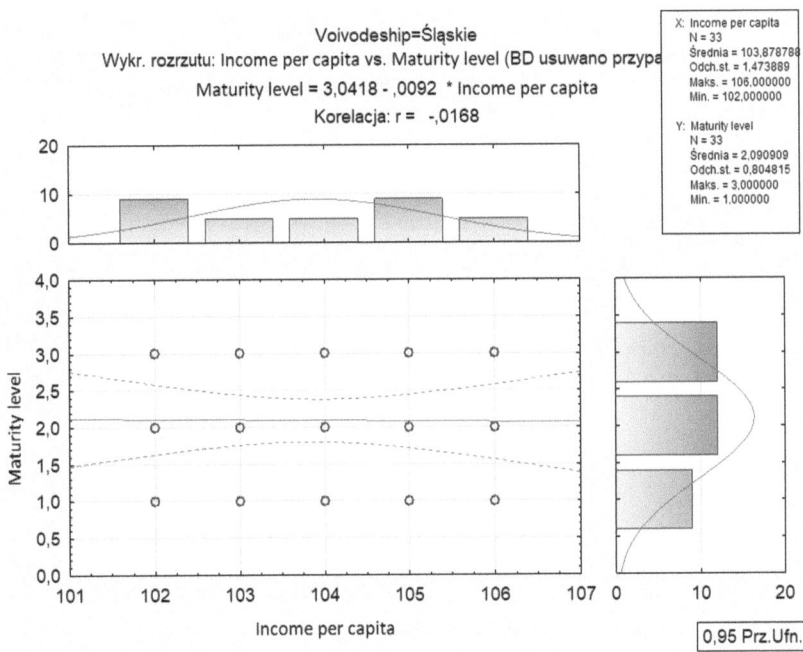

Fig. 6.28 Correlation between the amount of city income per capita in Śląskie Voivodeship and its level of logistics maturity. Source: own work

separated by a considerable distance from cities considered in theory and practice as fully intelligent entities. A higher rating of logistics maturity than SC maturity results from higher partial scores obtained in individual criteria, including the highest one relating to the level of cooperation between managers of the cities studied and stakeholders of city logistics, as well as the scope and time of undertaking activities related to city logistics. Nevertheless, it should be noted that in the real dimension of logistics maturity—included in the fourth criterion and referring to the level of implementation of intelligent solutions in the field of city logistics—the examined units fared the worst, which proves the relatively low effectiveness of the actions taken, despite the above-assessed cooperation with stakeholders and a better-evaluated scope (in terms of time and subject matter) of activities undertaken for the benefit of city logistics.

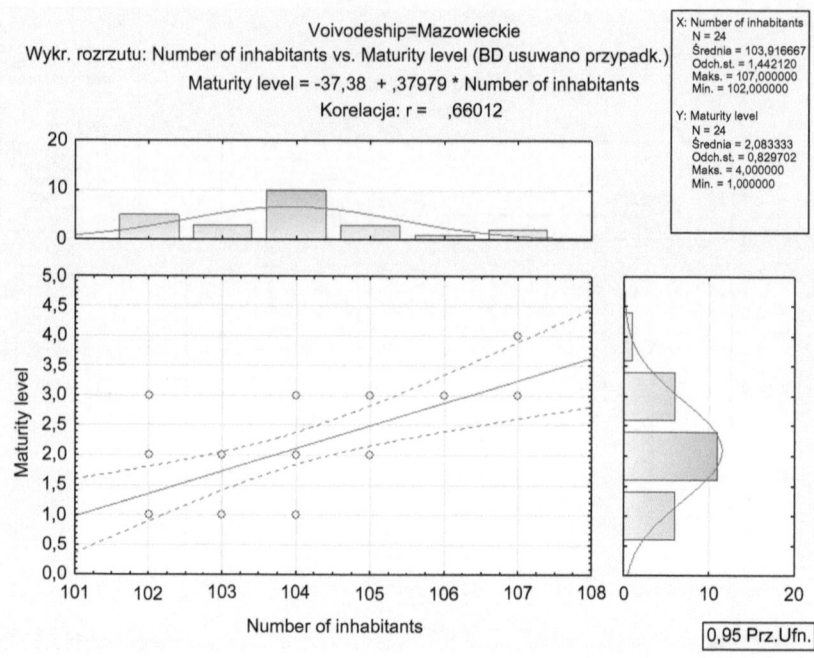

Fig. 6.29 Correlation between the size of cities in Mazowieckie Voivodeship and their level of logistics maturity. Source: own work

In the context of the data contained in Table 6.1, it is also worth emphasising that the advancement of Smart City solutions is characterised by a much higher variability than in the case of the assessment of logistics maturity, both holistically and individually at the level of individual criteria. Thus, the examined cities are more diversified in terms of city intelligence than logistic maturity, and therefore most of the studied cities are similar in terms of logistics, but in the case of the assessment of Smart City solutions, there are entities significantly deviating from the average, despite the low aggregate scores obtained in both major areas of analysis. This is also confirmed by the value of skewness for the SC rating, which proves the right-hand asymmetry of the distribution of ratings in this area. In the case of the logistics maturity assessment, the distribution is also right-handed, but less extreme. It is also more flattened (negative kurtosis) than is the case with the SC score, which indicates a more even

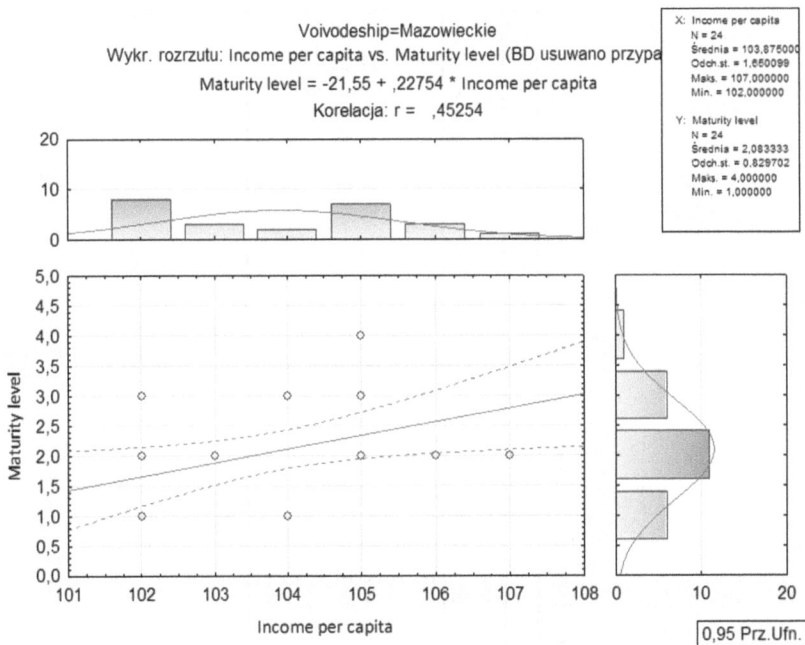

Fig. 6.30 Correlation between the amount of city income per capita in Mazowieckie Voivodeship and its level of logistics maturity. Source: own work

distribution of the grades awarded in this area. Therefore, it can be concluded that despite the low scores for SC and logistic maturity, the obtained results are characterised by a different distribution and variability.

In the second stage of the research, the surveyed cities were classified, taking into account both assessed aspects. The frequency of cities in particular groups is presented in Table 6.2, and the structure calculated on the basis of this frequency is presented in Table 6.3. In addition, the distribution of cities by the level of advancement of Smart City solutions at subsequent levels of logistics maturity is illustrated in Figs. 6.31, 6.32, and 6.33.

The analysed data show that at the first level of logistics maturity, there are mainly cities representing levels 0 and 1 of the advancement of intelligent city solutions. Additionally, the number of units at this level of logistic maturity decreased along with the increase in the value of the

Table 6.1 Values of basic statistics for the level of logistics maturity and the level of advancement of Smart City solutions

Variable	Parameters							
	Average	Median	Lower quartile	Top quartile	Standard deviation	Variability coefficient	Skewness	Kurtosis
Advancement of SC	1.27	1.00	0.00	2.00	1.38	108.75	1.27	1.01
Logistics maturity	1.86	2.00	1.00	2.00	0.79	42.56	0.39	-0.92
Criterion 1	2.64	3.00	2.00	3.00	0.99	37.66	0.10	-0.61
Criterion 2	2.93	3.00	2.00	4.00	1.09	37.19	-0.08	-0.75
Criterion 3	2.89	3.00	2.00	4.00	1.03	35.50	0.32	-0.70
Criterion 4	2.22	2.00	2.00	3.00	0.98	44.33	0.67	0.14

Source: own work

Table 6.2 Number of cities, taking into account the level of logistics maturity and the level of advancement of Smart City solutions

Logistics maturity level	The level of advancement of Smart City solutions						
	0	1	2	3	4	5	In total
1	44	42	12	5	0	4	**107**
2	44	37	12	12	2	2	**109**
3	7	23	7	13	1	10	**61**
4	1	0	0	0	1	1	**3**
5	0	0	0	0	0	0	**0**
In total	**96**	**102**	**31**	**30**	**4**	**17**	**280**

Source: own work

Table 6.3 Structure of cities, taking into account the level of logistics maturity and the level of advancement of Smart City solutions

Logistics maturity level	The level of advancement of Smart City solutions						
	0	1	2	3	4	5	In total
1	15.71%	15.00%	4.29%	1.79%	0.00%	1.43%	**38.21%**
2	15.71%	13.21%	4.29%	4.29%	0.71%	0.71%	**38.93%**
3	2.50%	8.21%	2.50%	4.64%	0.36%	3.57%	**21.79%**
4	0.36%	0.00%	0.00%	0.00%	0.36%	0.36%	**1.07%**
5	0.00%	0.00%	0.00%	0.00%	0.00%	0.00%	**0.00%**
In total	**34.29%**	**36.43%**	**11.07%**	**10.71%**	**1.43%**	**6.07%**	**100.00%**

Source: own work

assessment in the SC area. The situation was similar in the case of the second level of logistics maturity (Figs. 6.31 and 6.32).

At the third level of logistics maturity, the differentiation of cities in terms of the advancement of Smart City solutions was much greater (Fig. 6.33). This level was represented by the most numerous cities rated in the SC between 1 and 3, but this group also included the most numerous group of cities with the highest scores in terms of advancement in the implementation of Smart City solutions (ten entities).

The third level of logistics maturity was represented by only three cities, assessed in terms of SC as 0, 4, and 5. During the evaluation, cities at the fifth level of logistics maturity were not selected. Therefore, at the last two levels, no preliminary relationship can be found between the advancement level of Smart City solutions and the logistics maturity of the cities studied.

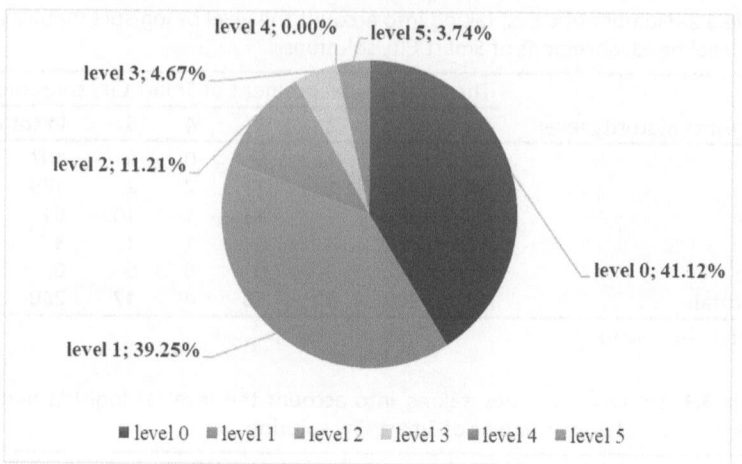

Fig. 6.31 The structure of cities according to the level of advancement of Smart City solutions at the first level of logistics maturity. Source: own work

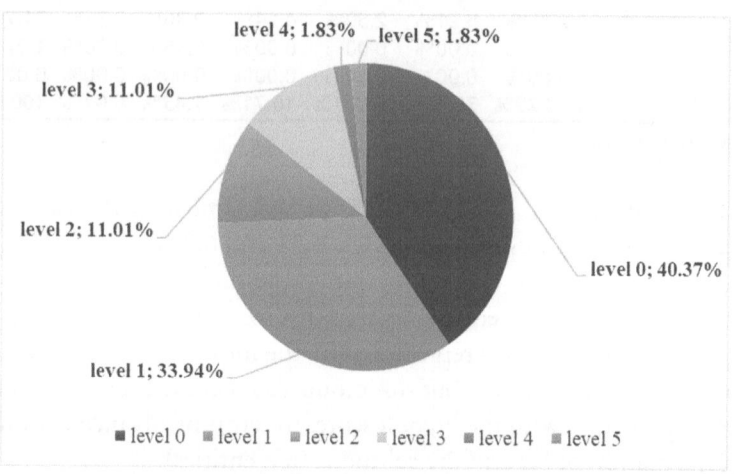

Fig. 6.32 The structure of cities according to the level of advancement of Smart City solutions at the second level of logistics maturity. Source: own work

To identify the links between the above-mentioned variables, in the third step, the Pearson's linear correlation coefficient was calculated. The results of these calculations are shown in Fig. 6.34.

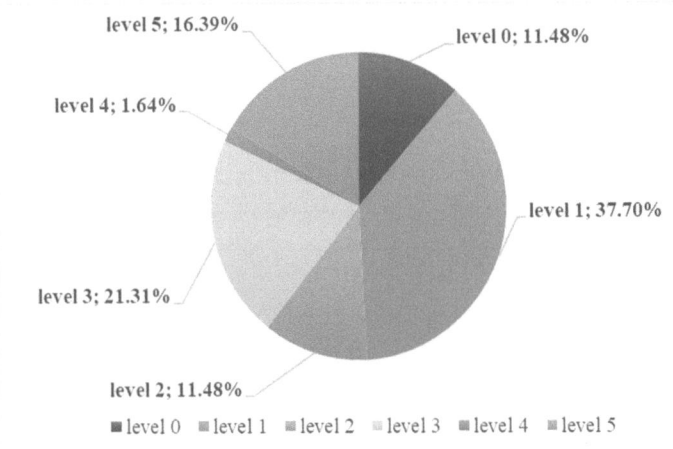

Fig. 6.33 The structure of cities according to the level of advancement of Smart City solutions at the third level of logistics maturity. Source: own work

The obtained values allow to conclude that between the logistic maturity and the advancement of city solutions there is a statistically significant and positive, but weak linear relationship ($r = 0.32752$). This means that as the advancement of Smart City solutions increases, the level of logistics maturity increases, but it is a slight increase—a change in the assessment of SC solutions—in accordance with the linear function shown in the figure—increases the level of logistics maturity by only 0.18722. The identified direction of the relationship confirms the dependence of city logistics and Smart City development, but the low value of the coefficient may result from the fact that, as indicated in chapter one, mobility aspects are only one of several determinants of Smart City development. The strength of the analysed relationship is also influenced by the differences in the distribution of the studied variables described in the first stage, even though their overall average rating is similar and low.

To detail the relationships between the studied criteria, in the fourth stage, the impact of individual criteria of logistic maturity (from 1 to 4) and the advancement of Smart City solutions on the final level of maturity of the studied cities were identified using multiple regression. The

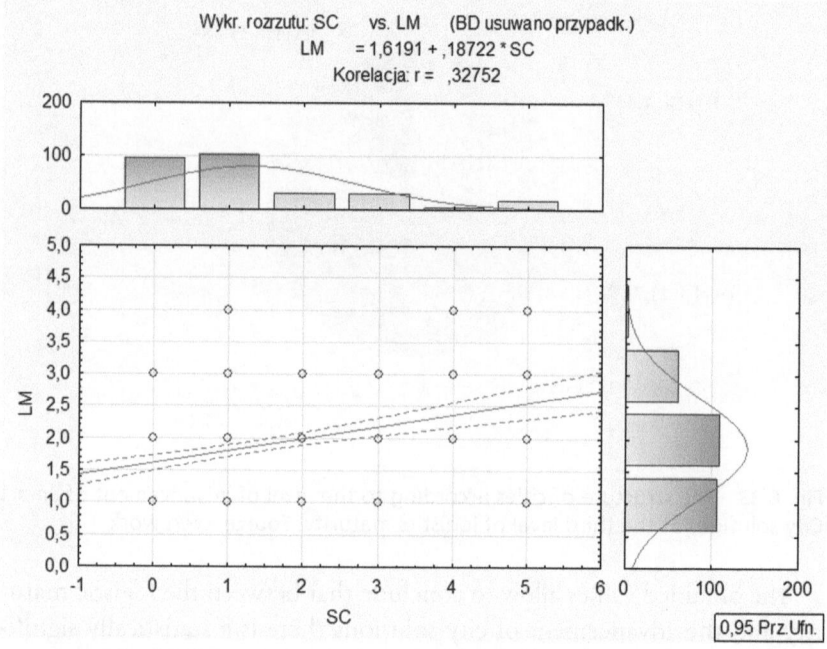

Fig. 6.34 Correlation between the level of advancement of Smart City solutions and the level of logistics maturity of the cities studied (statistically significant at $p < 0.05$). Source: own work

description of the functions is presented in two variants. The first model includes all analysed variables, that is:

- SC: level of advancement of Smart City solutions,
- criterion 1: the level of recognition of logistic areas in the formulated strategies of the analysed cities,
- criterion 2: the level of cooperation between the managers of the examined cities and the stakeholders of city logistics,
- criterion 3: the scope and timing of activities in the field of city logistics,
- criterion 4: the level of implementation of intelligent solutions in the field of city logistics.

and in the second, only those that turned out to be important for the described dependencies. The results of both analyses are presented in Tables 6.4 and 6.5, respectively.

The obtained results indicate a good fit of the model to the analysed variables (high values of the multiple determination coefficient and the improved multiple determination coefficient). Nevertheless, in the case of the function containing all examined parameters, the variables related to the advancement of Smart City solutions and related to criterion 3 (scope and time of taking actions in the field of city logistics) did not significantly contribute to the description of the level of logistics maturity, therefore they were removed from the second version of the multiple regression function.

Table 6.4 Multiple regression taking into account the level of advancement of Smart City solutions and all criteria for assessing logistics maturity

Variables	Dependent variable regression summary: $R = 0.86887$ $R^2 = 0.7550$ corr. $R^2 = 0.7505$ $F(5,274) = 168.85$					
	b^a	Std. er. with b^a	b	Std. er. with b	t(274)	p
Free term	–	–	-0.3005	0.0870	-3.4542	0.0006
SC	-0.0202	0.0329	-0.0116	0.0188	-0.6147	0.5392
Criterion 1	0.2716	0.0362	0.2163	0.0288	7.5030	0.0000
Criterion 2	0.2111	0.0341	0.1530	0.0247	6.1817	0.0000
Criterion 3	0.0474	0.0375	0.0365	0.0289	1.2648	0.2070
Criterion 4	0.5880	0.0378	0.4726	0.0303	15.5559	0.0000

Source: own work
[a]model with all analysed variables

Table 6.5 Multiple regression taking into account the level of advancement of Smart City solutions and significant criteria for assessing logistics maturity

Variables	Dependent variable regression summary: $R = 0.8679$ $R^2 = 0.7533$ corr. $R^2 = 0.7506$ $F(3,276) = 280.92$					
	b^a	Std. er. with b^a	b	Std. er. with b	t(274)	p
Free term	–	–	-0.2612	0.0818	-3.1932	0.0016
Criterion 1	0.2839	0.0348	0.2261	0.0277	8.1657	0.0000
Criterion 2	0.2139	0.0339	0.1550	0.0246	6.3092	0.0000
Criterion 3	0.5990	0.0330	0.4815	0.0265	18.1540	0.0000

Source: own work
[a]model with all analysed variables

Both in the first and second versions of multiple regression, the fourth criterion, which characterises the level of implementation of intelligent solutions in the field of city logistics, has the greatest impact on the level of logistics maturity of the analysed cities. If the rating of this criterion increases by 1, the level of logistics maturity of the city will increase by 0.5990. This means that real, documented implementations of Smart City solutions are of the greatest importance for the logistics development of cities. It is worth recalling here that during the survey, they included:

- intelligent waste management systems,
- solutions limiting freight road transport,
- solutions to reduce CO_2 emissions,
- noise reduction solutions,
- intelligent solutions for excluded people,
- intelligent passenger information systems,
- intelligent traffic control systems,
- low-mobility solutions,
- intelligent street lighting organisation,
- information about free parking spaces in the city in real time,
- technologies of accessibility and timeliness of public information,
- air quality monitoring technologies,
- modern geographic information systems,
- Internet of Things.

The second important determinant of the logistics maturity of the analysed cities is criterion 1: the level of including logistics areas in the formulated strategies of the analysed cities. An increase in this level by 1 causes an increase in the logistics maturity of the city by 0.2839. To a similar extent, the level of logistics maturity is determined by criterion 2: the level of cooperation between the managers of the cities studied and the stakeholders of city logistics. An increase in this level by 1 causes an increase in the logistics maturity of the city by 0.2139.

Based on the considerations and analyses undertaken, it appears that the studied Polish cities are characterised by a low level of both logistics maturity and the advancement of Smart City solutions. This is most likely due to two circumstances. The first is the financial problems of

Polish cities, mentioned in the first chapter, related to the low level of budget revenues and growing debt. The second is undoubtedly the relatively short period of building the Polish economy in free market conditions and its classification as a developing economy. Meanwhile, as shown by the results obtained, to improve logistics maturity, it is necessary, first of all, to implement the intelligent logistics solutions included in the fourth criterion for assessing logistics maturity. This means that Polish cities—wanting to follow the Smart City concept—will be forced, first of all, to provide financial and technological support for their urban infrastructure, which in the current economic and social conditions (the COVID-19 pandemic, the war in Ukraine) will be a very difficult and long-term challenge.

Conclusions

According to World Bank data today, some 55% of the world's population—4.2 billion inhabitants—live in cities. This trend is expected to continue. By 2050, with the urban population more than doubling its current size, nearly 7 of 10 people in the world will live in cities. With a growing population and consequently increasing congestion, cities have to cope with the negative effects of these phenomena. In the context of growing transport, organisational and social problems in cities, urban logistics and its tools are becoming increasingly important. When reflecting on the importance and future of urban logistics, priority should be given to its multidimensional and complex nature. For this reason, the authors have treated the issue of urban logistics in a comprehensive manner, analysing the existing theoretical output, and have also proposed a way to evaluate Smart City solutions. Based on a critical analysis of the literature and own research conducted in 280 Polish cities, the authors proposed an original approach to the study of the relationship between the level of application of intelligent solutions in the city and the level of logistic maturity of the city. Thus, it can be concluded that the conducted research made it possible to achieve the objectives of the monograph, to find answers to the formulated questions, and allowed for the formulation of general conclusions on the embedding of urban logistics in an intelligent, digital world:

- Factors building urban logistics maturity include the level of involvement of cargo transport stakeholders in the city, the formulation and implementation of urban logistics strategies (transport aspects in city strategies), cargo transport aspects in city strategic planning and integrated and sustainable passenger and cargo transport strategies in the city.
- The assessment of an urban logistics maturity can be made in relation to four criteria, which relate to the level of logistics maturity of a city:

  - level of inclusion of logistic areas in the formulated city strategies,
  - level of interaction between city managers and urban logistics stakeholders,
  - level of scope and timing of urban logistics activities,
  - level of implementation of smart urban logistics solutions.

- Two key themes of the literature relating to a city's level of progress in implementing the Smart City concept were used to assess the level of city intelligence:

  - theme of successive economic helixes: triple (business-city-science), quadruple (business-city-science-local community), and quintuple (business-city-science-local community-environmental organisations);
  - theme concerning the successive stages of Smart City development, covering the generation of cities referred to as: 1.0, 2.0, and 3.0.

- The basic relationships between a city's level of intelligence and its logistical maturity were made based on:

  - producing descriptive statistics for the aspects assessed in order to carry out a preliminary comparative analysis;

– classification of the examined cities taking into account the assessed aspects aimed at determining the structure and frequency of cities in particular groups;
– assessment of the correlation between the level of advancement of smart solutions and the level of logistical maturity of the studied cities, allowing the strength and direction of the relationship between the indicated variables to be identified;
– identification of the impact of individual logistics maturity criteria and the advancement of Smart City solutions on the final level of maturity of the surveyed cities.

• The research carried out indicated a strong correlation between innovation and logistical maturity of cities, manifested by the implementation of intelligent urban solutions. These include intelligent waste management systems, solutions to reduce vehicular cargo transport, solutions to reduce CO_2 emissions, solutions to reduce noise, intelligent solutions for excluded people, intelligent passenger information systems, intelligent traffic control systems, low mobility solutions, intelligent organisation of street lighting, real-time information about available parking spaces in the city, technologies for accessibility and timeliness of public information, air quality monitoring technologies, modern geographical information systems, and the Internet of Things.

The topicality of urban logistics issues and the growing importance of the challenges faced by city managers make this monograph useful for policy makers and city development strategists as well as for professionals involved in adapting urban space to the concept of sustainable development.

Index[1]

---

[1] Note: Page numbers followed by 'n' refer to notes.